APO NEW YORK 09454

life science investigations

man and the environment

life science investigations

man and the environment

Educational Research Council of America

Frederick A. Rasmussen

Paul Holobinko

Victor M. Showalter

HOUGHTON MIFFLIN COMPANY · BOSTON
New York · Atlanta · Geneva, Ill. · Dallas · Palo Alto

ACKNOWLEDGEMENTS

Many teachers, students, educators, and staff members of the Educational Research Council of America have contributed both directly and indirectly to the production of *Life Science Investigations*. It would be impossible to recall each contribution or pay due credit to all. However, we are especially obligated to the tireless efforts of Dr. Ted Andrews, Betty Schaffer Andrews, and Dr. James Joseph Gallagher, all of Governors' State University. Without their dedication and hard work the manuscript might never have evolved.

AUTHORS

Frederick A. Rasmussen, *Coordinator of the Life Science Writing Team*
Frederick A. Rasmussen is a Research Associate at the Educational Research Council of America and holds a B.S. degree in Zoology and a M.S. in Biology. He served in India as a Science Advisor to the United States Agency for International Development, and was on the writing team of both the Green and Yellow Versions of BSCS. He has over 13 years experience in secondary school teaching.

Paul Holobinko, *Assistant Coordinator*
Paul Holobinko is a Research Associate at the Educational Research Council of America and holds B.S. and M.S. degrees in Biology. He has over 10 years of secondary school teaching experience, including junior high school science, high school biology, and horticulture.

Victor M. Showalter, *Writer*
Victor M. Showalter is Director of Science at the Educational Research Council of America. He holds a B.S. degree in Chemistry, a M.S. in Chemistry and Physics, and a Ph.D. in Science Education. He has 16 years of teaching experience ranging from kindergarten to the graduate level. He has received both the STAR Award and the OHAUS Award from the National Science Teachers Association.

PRIMARY CONTRIBUTORS

JOHN DAY, Teacher-Writer, Camelback High School, Phoenix, Arizona
JOAN FELMER, Teacher-Writer, North Olmsted Junior High School, North Olmsted, Ohio
ROBERT HOWE, Writer, Ohio State University, Columbus, Ohio
WILLARD KORTH, Research Design, Evaluation, University of Pittsburgh, Pittsburgh, Pennsylvania
RICHARD M. PEARCE, Teacher-Writer, Skagit Valley College, Mount Vernon, Oregon
VIVIEN RAUNEGGER, Illustrator, Educational Research Council of America
KEN SHIPLEY, Artist-Designer, Educational Research Council of America
ROBERT SMITH, Teacher-Writer, Harding Junior High School, Lakewood, Ohio
TED SURDY, Biologist-Writer, Southwest Minnesota State College, Marshall, Minnesota
JOSEPH WILLIS, Teacher-Writer, Yuba City High School, Yuba City, California

CONSULTANTS TO THE ERC LIFE SCIENCE PROGRAM

DR. GARLAND E. ALLEN, History and Philosophy of Science, Washington University
DR. WILLIAM W. HAMBLETON, Geology, University of Kansas
DR. PAUL DEHART HURD, Science Education, Stanford University
DR. L. CARROLL KING, Chemistry, Northwestern University
DR. ADDISON E. LEE, Science Education and Biology, University of Texas
DR. JOSEPH D. NOVAK, Science Education, Cornell University
DR. CLIFFORD SWARTZ, Physics, State University of New York, Stony Brook
DR. FLETCHER G. WATSON, Science Education, Harvard University
DR. HARVEY BENDER, Biology, Notre Dame University

CONTRIBUTORS

ANN ANDERSON, North Junior High School, Brockton, Massachusetts
DON BAESEL, Emerson Junior High School, Lakewood, Ohio
BARBARA BARRETT, Tower Heights Middle School, Centerville, Ohio
HARLEY BEAL, Harding Junior High School, Lakewood, Ohio

COPYRIGHT © 1971 BY EDUCATIONAL RESEARCH COUNCIL OF AMERICA
All rights reserved. No part of this work may be reproduced or transmitted in any form or by any means, electronic or mechanical, including photocopying and recording, or by any information storage or retrieval system, without permission in writing obtained from Houghton Mifflin Company, 110 Tremont Street, Boston, Massachusetts 02107. Printed in the U.S.A.

Library of Congress Catalog Card Number: 71-131290

Student's Edition ISBN: 0-395-10869-1
Teacher's Edition ISBN: 0-395-10870-5

GERALD BEALS, West Junior High School, Brockton, Massachusetts
KRIS BENNETT, Bay Middle School, Bay Village, Ohio
PEARL BIGGIN, School Street Junior High School, Bradford, Pennsylvania
HERBERT BLAKANN, Ford Junior High School, Berea School District, Brookpark, Ohio
VIRGINIA BODDY, Ford Junior High School, Berea School District, Brookpark, Ohio
LYNN BOHN, Falls Middle School, Olmsted Falls, Ohio
WILLIAM BROWN, Elyria Public Schools, Elyria, Ohio
SYLVIA CALE, Hithergreen Middle School, Centerville, Ohio
SUSAN CASTLE, North Olmsted Junior High School, North Olmsted, Ohio
KEN CEROKY, Chardon Middle School, Chardon, Ohio
DON CLARICO, Middleburg Heights Junior High School, Middleburg Heights, Ohio
ELIZABETH COBB, West Junior High School, Brockton, Massachusetts
VIVIEN CROWLEY, East Junior High School, Brockton, Massachusetts
J. PETER DAHLBORG, West Junior High School, Brockton, Massachusetts
ANNA DAILEY, North Junior High School, Brockton, Massachusetts
SHIRLEY DEUVALL, Ford Junior High School, Cleveland, Ohio
JOHN DUGDALE, East Junior High School, Brockton, Massachusetts
DAVID DUVALL, Middleburg Heights Junior High School, Middleburg Heights, Ohio
JOHN EVANS, Falls Middle School, Olmsted Falls, Ohio
FRED FANNING, Hithergreen Middle School, Centerville, Ohio
JOHN FRITSCHE, Messiah Lutheran School, Fairview Park, Ohio
THOMAS FROMMER, Messiah Lutheran School, Fairview Park, Ohio
ARTHUR GERVAIS, West Junior High School, Brockton, Massachusetts
LARRY HALL, Ballard Junior High School, Niles, Michigan
JOHN HENRY, Roehm Junior High School, Berea, Ohio
CHARLOTTE HUDGIN, Summit Junior High School, Summit, New Jersey
HARRY HUSAK, North Olmsted Junior High School, North Olmsted, Ohio
EDMUND JANETZKE, Bethlehem Lutheran School, Euclid, Ohio
JOHN KARO, South Junior High School, Brockton, Massachusetts
KENNA KELLY, North Junior High School, Brockton, Massachusetts
PATRICIA KOTNIK, Harding Junior High School, Lakewood, Ohio
MARY LANDES, Bay Middle School, Bay Village, Ohio
DENNIS LEONARD, Bay Middle School, Bay Village, Ohio
ANTHONY LIOTTA, Mayfield Junior High School, Mayfield, Ohio
CHRISTOPHER LOWRIE, West Junior High School, Brockton, Massachusetts
WILLIAM MAGRAME, Ballard Junior High School, Niles, Michigan
ROBERT MAHOUSKY, Harding Junior High School, Lakewood, Ohio
LESLIE MALMGREN, North Junior High School, Brockton, Massachusetts
ANN MASSA, North Olmsted Junior High School, North Olmsted, Ohio
ALICE MCCARTHY, Ford Junior High School, Berea School District, Brookpark, Ohio
NANCY MCCUNE, Hithergreen Middle School, Centerville, Ohio
SAM MCCUTCHEON, Floyd C. Fretz Junior High School, Bradford, Pennsylvania
FRANK MOORE, Horace Mann Junior High School, Lakewood, Ohio
JOHN MURPHY, Falls Middle School, Olmsted Falls, Ohio
DON MYERS, Tower Heights Middle School, Centerville, Ohio
LOUIS NIELSON, Ballard Junior High School, Niles, Michigan
LEE NIHAN, South Junior High School, Brockton, Massachusetts
RICHARD PACE, Chardon Middle School, Chardon, Ohio
JOSEPH PAGE, Falls Middle School, Olmsted Falls, Ohio
EDWARD PILARSKI, Ring Lardner Junior High School, Niles, Michigan
CONCETTA REAVIS, North Junior High School, Brockton, Massachusetts
WILLIAM RUSCH, St. Thomas Lutheran School, Rocky River, Ohio
RICHARD SCHINKEL, Ring Lardner Junior High School, Niles, Michigan
WILLIAM SCHWENCK, Summit Junior High School, Summit, New Jersey
GARY SINCK, Hithergreen Middle School, Centerville, Ohio
DIANE SMITH, Cuyahoga Heights High School, Cuyahoga Heights, Ohio
FREDERICK SMITH, School Street Junior High School, Bradford, Pennsylvania
VIRGINIA SMITH, West Junior High School, Brockton, Massachusetts
TED SOBIERAJI, Independence Middle School, Independence, Ohio
GORDON SPENCER, Hithergreen Middle School, Centerville, Ohio
ARTHUR STEHLIK, Emerson Junior High School, Lakewood, Ohio
HAROLD STODDARD, Summit Junior High School, Summit, New Jersey
SANDRA SURFACE, Hithergreen Middle School, Centerville, Ohio
SANDRA SWARD, Summit Junior High School, Summit, New Jersey
JANET THOBABEN, Tower Heights Middle School, Centerville, Ohio
CAROLYN THOMES, Tower Heights Middle School, Centerville, Ohio
MARY TOOMEY, South Junior High School, Brockton, Mass.
ROSE VERNON, Tower Heights Middle School, Centerville, Ohio
JOEL WELLS, Cuyahoga Heights Junior High School, Cuyahoga Heights, Ohio
JANICE WILLIAMS, Harding Junior High School, Lakewood, Ohio
NADINE WOLLENZIER, Emerson Junior High School, Lakewood, Ohio

Preface to the teacher

Life Science Investigations began in 1965 as part of a general effort to improve K-12 science teaching. A group of science teachers and administrators from schools affiliated with the Educational Research Council of America met with a committee of consultants, and evolved a plan for a new middle school life science program. This plan called for the life science course to:

a. Emphasize an inquiry approach to human-oriented problems in biology,
b. Analyze similarities and differences between man and other organisms,
c. Investigate the interaction of man and his environment, and
d. Develop a perspective on some of the major biological problems facing man.

The First Experimental Edition was prepared and used in classrooms during the 1966–67 school year. Based on the recommendations of teachers, consultants, and the ERC science staff, plans were made for extensively reorganizing this edition. The Second Experimental Edition was subsequently written during the summer of 1967 and early months of 1968. Over 100 teachers and 16,000 students in schools affiliated with the Educational Research Council of America used and evaluated the Second Experimental Edition. Finally, when sufficient feedback had been gathered, plans were made to write a third, definitive edition.

So, in effect, *Life Science Investigations* represents almost five years of careful and continuous writing, field-testing, and revising. The end product is this new and exciting course in life science.

The Educational Research Council of America is a nonprofit corporation dedicated to developing exemplary curricular materials for grades K-12 in all subject areas for all students. The Council is composed of a central staff of about 200 and the 30 affiliated school systems. Taken together, they form a team for developing relevant, innovative, and effective instructional materials.

Life Science Investigations is only one of several courses that have been developed in this framework. Other courses for different grade levels are currently being devised, field tested, and revised.

FREDERICK A. RASMUSSEN, *Coordinator*

Preface to the student

You will find that this book is different from most school books you have seen. It will help you to investigate scientifically some of your own questions about living things.

You will be learning about life science this year through your own activities. This may be one of the most enjoyable and rewarding courses you have ever had because:

You will spend most of your time investigating life science rather than reading about it.

You will be working with a small group of classmates on common problems.

Your teacher will be a helper and advisor rather than just a source of information.

When you have completed this course you will have learned to do many things. Some of these are:

a. Setting up experiments to test your ideas about the behavior of plants and animals, including yourself,
b. Finding out what conditions affect living things,
c. Discovering how *you* can affect living things, and
d. Planning different ways to solve the problems that face people living on this planet.

FREDERICK A. RASMUSSEN, *Coordinator*

Guide to pronunciation

a, like *a* in p*a*t	y, ye, eye, like *ie* in p*ie*	uh, like *a* in *a*bout
ah, like *a* in f*a*ther	i, ih, like *i* in p*i*t	ur, like *ure* in fut*ure*
ai, like *a* in c*a*re	o, like *o* in p*o*t	g, like *g* in *g*a*g*
ay, like *ay* in p*ay*	oo, like *oo* in t*oo*k	j, like *j*, *dg* in *j*u*dg*e
aw, like *aw* in p*aw*	oh, like *oe* in t*oe*	n, like *n*, *en* in *n*o, sudd*en*
e, eh, like *e* in p*e*t	oi, like *oi* in n*oi*se	ng, like *ng* in the thi*ng*
ea, like *ea* in f*ea*r	ow, like *ou* in *ou*t	s, like *s*, *c* in *s*au*c*e
ee, like *e* in b*e*	u, uh, like *u* in c*u*t	z, like *z* in *z*ebra
ew, like *oo* in b*oo*t	er, like *ur* in *ur*ge	zh, like *si* in vi*si*on

Contents

UNIT ONE — INVESTIGATING LIVING THINGS — xii

Investigation 1 — Inquiring — 2

PROBLEM 1-1 What is a good question? — 3
PROBLEM 1-2 Asking questions about people — 7
MASTERY ITEM 1-1 How can this be? — 9

Investigation 2 — Making scientific observations — 12

PROBLEM 2-1 What can you observe about people? — 14
PROBLEM 2-2 What can you infer from a person's hands? — 15
PROBLEM 2-3 How much can you learn by observing living things? — 18
PROBLEM 2-4 How fast does your heart beat? — 21
PROBLEM 2-5 How often do people blink? — 26
MASTERY ITEM 2-1 Describing an animal — 29
MASTERY ITEM 2-2 Interpreting a photograph — 29
MASTERY ITEM 2-3 The truth about two trees — 30
MASTERY ITEM 2-4 An inference about fish — 31
MASTERY ITEM 2-5 How fast do people breathe? — 33

Investigation 3 — Perceiving the world — 34

PROBLEM 3-1 How do animals get information? — 36
PROBLEM 3-2 How reliable are your senses? — 42
PROBLEM 3-3 How does experience affect perception? — 50
MASTERY ITEM 3-1 Perceiving a city street — 54
MASTERY ITEM 3-2 Which way is up? — 55
MASTERY ITEM 3-3 A perception experiment — 56
MASTERY ITEM 3-4 A real flying saucer? — 57

Investigation 4 — Organizing information — 58

PROBLEM 4-1 How can things be classified? — 60
PROBLEM 4-2 Making hypotheses—why do different people have different heartbeat rates? — 65
MASTERY ITEM 4-1 How many ways can you classify these students? — 72
MASTERY ITEM 4-2 Which birds eat the same foods? — 75
MASTERY ITEM 4-3 Why do some animals' hearts beat faster than others? — 75
MASTERY ITEM 4-4 Who can run fifty yards the fastest? — 76

Investigation 5 — Describing living things — 78

PROBLEM 5-1 What does "alive" mean? — 81
MASTERY ITEM 5-1 Living or nonliving? — 84
MASTERY ITEM 5-2 Plant or animal? — 84

Investigation 6	**Designing experiments**	86
	PROBLEM 6-1 Do aquatic plants and animals exchange gases?	88
	PROBLEM 6-2 Measuring gas exchange	91
	MASTERY ITEM 6-1 A survival experiment	96
	MASTERY ITEM 6-2 Do roots need oxygen?	97
Investigation 7	**Investigating behavior**	98
	PROBLEM 7-1 The goldfish puzzle	100
	MASTERY ITEM 7-1 What makes flies thirsty?	103
	MASTERY ITEM 7-2 What factors affect your pulse rate?	104
	MASTERY ITEM 7-3 How does exercise affect your pulse rate?	106
Investigation 8	**Learning**	108
	PROBLEM 8-1 The finger maze	111
	PROBLEM 8-2 The mouse in the maze,™ a simulated experiment	114
	PROBLEM 8-3 Can animals learn to escape?	118
	MASTERY ITEM 8-1 Studying with the radio on	121
	MASTERY ITEM 8-2 Learning without reward	121
	MASTERY ITEM 8-3 Can ants learn?	123

UNIT TWO

THE ENVIRONMENT AFFECTS LIVING THINGS

124

Investigation 9	**How does light affect euglenas?**	126
	PROBLEM 9-1 How do euglenas react to light and dark?	129
	PROBLEM 9-2 Are other light conditions important?	131
	MASTERY ITEM 9-1 Where do euglenas live in this pond?	132
	MASTERY ITEM 9-2 The sun and ocean-dwelling organisms	133
Investigation 10	**What influences seed germination?**	134
	PROBLEM 10-1 How do seeds respond to water?	137
	PROBLEM 10-2 How does temperature affect germination?	142
	PROBLEM 10-3 Do seeds respond to light?	145
	MASTERY ITEM 10-1 The best way to germinate seeds	151
	MASTERY ITEM 10-2 What makes peppergrass sprout?	151
Investigation 11	**The water needs of plants**	154
	PROBLEM 11-1 Why do plants wilt?	155
	PROBLEM 11-2 What increases water loss in plants?	158
	PROBLEM 11-3 Does leaf area affect water loss?	164
	MASTERY ITEM 11-1 What are greenhouses for?	166
	MASTERY ITEM 11-2 Adapting to climate	167
Investigation 12	**What environment suits a sow bug?**	168
	PROBLEM 12-1 What makes a sow bug react?	171
	MASTERY ITEM 12-1 Where will the blowflies settle?	174
	MASTERY ITEM 12-2 The survival value of animal behavior	176

Investigation 13	The energy needs of a community	178
	PROBLEM 13-1 A community in the dark	179
	MASTERY ITEM 13-1 Life in a pond	188
	MASTERY ITEM 13-2 Life on a coral reef	190

UNIT THREE — LIVING THINGS AFFECT EACH OTHER   192

Investigation 14	Coyotes and their prey	194
	PROBLEM 14-1 The coyote problem	195
	MASTERY ITEM 14-1 Managing a lake	205
	MASTERY ITEM 14-2 A case history in managing wildlife	207
Investigation 15	Competing with microbes	210
	PROBLEM 15-1 Why does food rot?	211
	PROBLEM 15-2 Do microbes cost you money?	213
	PROBLEM 15-3 What happens to garbage and trash?	216
	MASTERY ITEM 15-1 A bulging can on the shelf	217
	MASTERY ITEM 15-2 Eating a woolly mammoth	218
	MASTERY ITEM 15-3 Do we need a "super-preservative?"	219
Investigation 16	Planet management	220
	PROBLEM 16-1 How can Clarion be improved?	222
	MASTERY ITEM 16-1 The Earth Management Game	232
	MASTERY ITEM 16-2 Improving The Planet Management Game™	233
	MASTERY ITEM 16-3 Planet management and Earth today	233
Investigation 17	Organism versus organism	234
	PROBLEM 17-1 Offense and defense	235
	MASTERY ITEM 17-1 When the tide turns red	243
	MASTERY ITEM 17-2 Turnips and marigolds	244
Investigation 18	The Redwood Controversy™	246
	PROBLEM 18-1 What will be the fate of the redwoods?	247

UNIT FOUR — MAN AFFECTS THE ENVIRONMENT   252

Investigation 19	Who pollutes your environment?	254
	PROBLEM 19-1 How do you affect your environment?	255
	MASTERY ITEM 19-1 Family pollution	262
	MASTERY ITEM 19-2 The dandelion problem	262

Investigation 20		**Is Lake Erie dead?**	264
	PROBLEM 20-1	What's wrong with Lake Erie?	265
	PROBLEM 20-2	Can Lake Erie be helped?	281
	MASTERY ITEM 20-1	Which lake is more polluted?	281
	MASTERY ITEM 20-2	Is Crystal Lake dying?	282
	MASTERY ITEM 20-3	Can Crystal Lake be saved?	284
Investigation 21		**What is the price of progress?**	286
	PROBLEM 21-1	How did the rivers get hot?— The Thermal Pollution Game	288
	PROBLEM 21-2	The atomic power plant controversy	302
	MASTERY ITEM 21-1	How will you invest your money now?	324
Investigation 22		**What can we do about pests?**	326
	PROBLEM 22-1	Controlling pests	328
	MASTERY ITEM 22-1	A pesticide poll	350
Investigation 23		**Why is clean air important?**	352
	PROBLEM 23-1	How much pollution falls on your house?	354
	PROBLEM 23-2	What do airborne particles look like?	358
	PROBLEM 23-3	How much dirt do you inhale?	362
	PROBLEM 23-4	Can air pollution harm people?	366
	PROBLEM 23-5	How does sulfur dioxide affect plants?	371
	MASTERY ITEM 23-1	Planning a local pollution study	373
	MASTERY ITEM 23-2	Inhaling pollution	373
	MASTERY ITEM 23-3	Do automobile exhausts harm corn?	376
	MASTERY ITEM 23-4	Does smoking cause lung cancer?	376
	MASTERY ITEM 23-5	Stop smoking!	379
Investigation 24		**The pollution game**	380

Appendix

PART A	Observing with the microscope	388
PART B	How to graph data	396
PART C	Fahrenheit and Celsius thermometers	407

Glossary   408

Credits   411

Index   414

Unit one

Investigating living things

Questions: How fast does your heart beat? How are living things unlike nonliving things? Can all animals learn? Why do people differ in their tastes and attitudes? Unit One in this book was planned to help you investigate questions like these.

How do you get answers to questions about yourself and other living things? One way is by inquiring for yourself, like a scientist. In this unit you will be able to develop and practice some important skills for investigators. The activities in the unit should help you answer questions and make decisions not just in school, but out of school also.

If Unit One is successful, you will learn to do these things and some others:

a. Ask interesting questions that you can investigate yourself.
b. Predict possible outcomes of your investigations.
c. Set up simple experiments to study plant and animal behavior.
d. Explain how different living things learn about the world around them.

Check this list after you do the activities in the unit. See if you know more about how to investigate living things.

FIGURE 1.1
What kinds of questions can be investigated?

© 1961 Saul Steinberg:
From *The New World* (Harper & Row)
Originally in The New Yorker.

Investigation

1 Inquiring

Do I have the right answer? What are we supposed to do? Why? Questions! How many have you asked today? How often do you try to find your own answers?

This Investigation is about *asking* questions. Its purpose is to help you ask questions that you can investigate for yourself.

Problem 1-1

What is a good question?

People ask questions because they want information. But some questions are easier to investigate than others. These questions help you get the information you want. Do questions that can be investigated share some characteristics? Here is a way to find out.

Materials
tracing paper   ruler
pencil

Gathering data
Data (DAY-tuh) is information. To help answer questions or solve problems, you gather data. In this Problem you will

FIGURE 1.2
What do you see when parallel lines are placed over this figure?

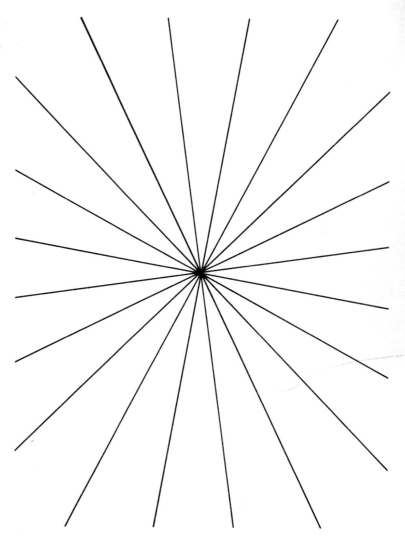

try to find out what kinds of questions can be investigated most easily. A person usually follows a procedure or plan to gather data. For this Problem, follow the procedure given below.

Using a ruler, draw two parallel lines five inches long and exactly three-quarters of an inch apart on the tracing paper. Place your paper over Figure 1.2. Center your lines between the side lines of the drawing. What do you see?

Inquiring

To understand what you see, you probably need more information. There are questions about Figure 1.2 after this paragraph. They may help you to find some useful information. Read these questions and try to answer them. Use the materials provided. Do not ask for help from your teacher or classmates. This is *not* a test, and the results will not be graded. The idea is simply to find out what kinds of questions can be investigated.

QUESTIONS ABOUT FIGURE 1.2
a. Are the two "curved" lines really curved?
b. Why do I see the lines as curved?
c. How long are the "curved" lines?
d. Do other people see the two lines as curved?
e. What would the two "curved" lines look like if half as many lines crossed in the center of the drawing? Twice as many?
f. Does your brain make you see the two lines as curved?
g. What is this drawing supposed to mean?
h. What would the two "curved" lines look like if they were slowly moved across the drawing from left to right or from top to bottom?
i. Do the two lines look curved when you sight along the Figure from one end instead of looking straight at it?
j. Do the two lines still look curved when they are moved closer together or farther apart?

Recording data

In your notebook list your answers to the questions about Figure 1.2. This is a record of your data. You will refer to it later.

Analyzing data

After gathering data, you study each bit of it to see what it tells you. To **analyze** (AN-uh-lyz) means to study the parts of something carefully. You hope your data will help solve the problem you are investigating.

The questions below will help you analyze your data. Discuss them with your classmates and teacher.

a. Do you now have any idea why the lines appear curved? If you do, which questions led you to this information?

b. Do these helpful questions share any characteristics?
c. Which questions, if any, couldn't you answer? What characteristics made these questions too difficult to answer?
d. Did any questions lead to information that did *not* help you understand the "curved" lines?
e. Have you thought of any new questions that might lead to more useful information? What characteristics do these questions have?

By now you should have learned some characteristics of questions that you can investigate. List these characteristics in your notebook. Many times during *Life Science Investigations,* you will ask questions and then answer them yourself. Use this list as a model or guide to help you ask good questions.

Problem 1-2

Asking questions about people

You already know quite a bit about people. But you may not be able to explain the two activities in this Problem. Try out the activities. Then, by asking questions and answering them, you can learn about yourself and other people. Remember to think of questions that help you seek information for yourself.

Materials

list of "memory words"   pencil
"test paper"
Ask your teacher for any other materials you need.

Gathering data

Read through the two parts of this Problem. Do both activities. Which one interests you more? You can investigate either one or both of them.

MEMORY WORDS

For this activity the class separates into two groups. Your teacher will place a list of words face down on your desk. Do not turn the list over until the teacher says so.

You have one minute to memorize the list of words. Then turn the list face down again. Write all the words you can remember on another sheet of paper. Write the words in the same order as they are listed. You have one minute to write.

Next check to see how many words you remembered correctly. Write the number on your paper. Two or three people in your group can add the numbers to find your group's score. Compare your group's score with the other group's score.

TEST PAPER

Take a piece of "test paper" and lick it. Your teacher will do this, too. Examine the paper after about a minute. Then compare it with your classmates' and teacher's test papers.

Can you explain the results of the activity you did, or do you need more information? What questions will help you seek the information you need?

First, write down what the problem is. It helps to say just what the problem is before you ask more questions about it. Now make a list of as many questions as you can about the activity. You want questions that might help you find out what is going on. Then try to find answers for your questions.

Recording data

Write down the problem and list your questions in your notebook, as you think of them. Leave a space after each question for writing an answer.

Analyzing data

There are two things to analyze now. One is the outcome of your activity. The second thing is the list of questions you asked. In Problem 1-1 you looked for questions that could be investigated. You can do the same thing now to analyze your own list of questions and answers. First find the questions that helped you get useful information. Then find the characteristics these questions share.

Some of your classmates asked about the same activity you did. Discuss your questions and answers with them. Here are some suggestions to help you get a discussion started.

a. Did the answers to your questions help you understand the activity?
b. Were there questions you could not investigate? Can you change these questions so they can be investigated?
c. Did you ask any questions about the entire activity instead of part of it? Do you think these were good questions?
d. Did you ask any questions that did not need specific answers such as "yes" or "no" or an amount? Do you think these were good questions?
e. Were there any questions you could not answer, but still think are good questions? How are these questions similar to the ones you *could* answer?

Inquiring

FIGURE 1.3
When you lick the paper, be sure to get it quite wet.

Mastery Item 1-1

How can this be?

There are Mastery Items at the end of each Investigation. They will check how well you have learned the skills and ideas in an Investigation. They will show you where you need more practice or help. Then your teacher can suggest ways to help you improve. The Mastery Items are *not* tests for determining your grade for the course. They are a method for analyzing your own progress. First find your own answers. Then compare them with the answers printed upside down at the end of the Items.

10   Inquiring

Here is the first Mastery Item. Questions you can try to answer yourself are usually about specific parts of problems or situations. Has this Investigation helped you to ask questions you can find answers for? This Mastery Item will tell you.

Examine the situation in Figure 1.4. Write as many questions as you can to help you explain what is happening. Imagine that you are at the scene in the picture, trying to

FIGURE 1.4
What is happening here?

answer your questions. The people in the picture will talk to you, but they can't explain what is happening. List your questions in your notebook.

Key

You are successful in asking questions you can investigate, if:

a. You asked at least five questions.
b. At least three of your questions asked for specific answers, such as "yes," or "no," a number, or an amount.
c. At least three questions asked about specific and unusual parts of the situation.
d. At least four of your questions could be investigated.
e. You did not expect more than one question to be answered by your teacher.

FIGURE 1.5
Some questions just won't leave you alone.

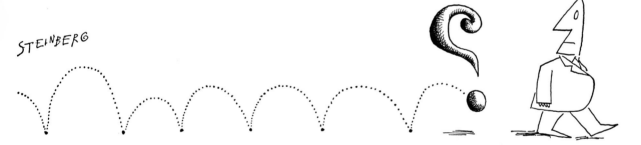

© 1961 Saul Steinberg.
From *The New World* (Harper & Row)
Originally in The New Yorker.

FIGURE 2.1
Are these two hamsters the same?

Investigation

2 Making scientific observations

How alert are you? Do you really notice all the things that happen around you? Do all trees look the same, all scrambled eggs taste the same, all car engines sound the same? Or can you see differences among the leaves on the same tree?

Did y*o*u notice that all the letter*s* in this sentence aren't the same type face?

In this Investigation you will develop your powers of observation. You will also learn to make observations especially useful in science.

You have probably seen the people in your class many times. But have you ever watched any one of them carefully and deliberately? Have you ever observed one of your classmates scientifically? In this Problem you will.

FIGURE 2.2
Are these two hands the same?

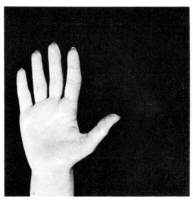 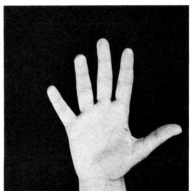

Problem 2-1

What can you observe about people?

Gathering data

Pick out a person sitting near you in the classroom. He or she will be your subject for thorough observation. Observe your subject very carefully for 10–15 minutes. Write a short note for each observation that you make.

It might be fun to imagine that you and your classmates are visiting scientists from Mars. Pretend that you are on Earth to study humans. Before your spaceship leaves, you have a short time to learn all you can about Earthlings by observing one subject. You can imagine that the other scientists from Mars (your classmates) are doing the same thing with other subjects. Later on you will have a chance to compare observations. All together, you should be able to discover many characteristics of Earthlings.

FIGURE 2.3
What can you observe about this student?

Recording data

Make a numbered list of your observations in your notebook. You don't have much time, so make your notes short. But make them neat, so other people can read them easily.

Analyzing data

Compare your observations with your classmates'. Did almost everyone observe the same things, like hair color and height? Did anyone group his observations into reports on the subject's appearance and reports on the subject's actions?

You can learn more about making observations if you discuss these questions with your classmates and teacher.

a. Which written observations are clear in their meaning? Do some give an incorrect picture? A good test of clearness is to imagine reading your observations to a friend over the telephone. Would he understand you?
b. Are all observations based on sight alone? Or were other senses like hearing used?
c. What value is there in making observations in a planned, orderly way? How could you improve the order of your observations?
d. Have some of your notes described more than you actually observed?

Problem 2-2

What can you infer from a person's hands?

The word **infer** (in-FUR) may be new to you. It is very important, especially in science. When someone infers, he goes beyond simple observation. He imagines something that may or may not be true. For instance, if you went outside and observed that the sidewalks were wet, you might infer that it had been raining. "Wet sidewalks" is an observation. It can be seen directly. However, "raining" is an inference. It was not observed directly. In fact, someone might have sprayed the sidewalks with a hose, or maybe a water pipe broke. If you use your imagination, you might think of hundreds of explanations for a wet sidewalk.

FIGURE 2.4
Pick out one person in this crowd. What can you infer about him or her?

Here's another example that compares an observation to an inference. Imagine that the school's fire bell rings during a class. You hear the bell clanging. That is an observation. If you say to yourself, "Another practice drill!" that is an inference. It may be the right reason for the fire bell's ringing. But there may be other reasons—like a real fire. In that case you will observe other things as you leave the building. These will tell you whether your original inference was correct.

Not all inferences are easy to check directly. If you saw six large oak trees growing in a straight line, you might infer how they got that way. How could you check your inference?

Gathering data

Select one of your classmates as the subject for your study. Observe your subject's hands carefully. Try to make inferences about the person on the basis of your observations of the hands only.

Making scientific observations

For example, by observing the shape of the fingers, can you infer the person's height or weight? From your observations of the skin color, what can you infer about hair color? Try not to look at the rest of the person while you observe his hands.

Spend at least 10 minutes making observations and inferences. When you have a list of observations and inferences, check to see how well they go together.

Your teacher will arrange for everyone to observe the hands of someone from outside the class. You will not be able to see the whole person, just the hands. What can you infer about the person? Is the person male or female? How old is the person? How tall? What color hair? What kind of work does he or she do? Don't show anyone your inferences yet.

FIGURE 2.5
Which hands belong to which person?

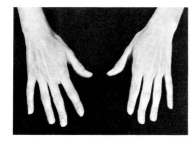
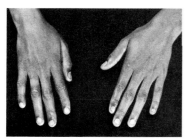
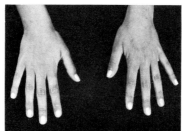
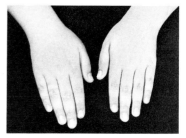

Recording data

Divide a notebook page into two columns. Head one column "Observations," the other "Inferences." In the first column, describe each observation. In the other column, describe the inference you made from the observation. One observation may lead to more than one inference. Or several observations may lead to the same inference.

Keep the data on your classmate and on the unknown subject separate.

Analyzing data

Making inferences is a skill. You can improve your skill by practice and by discussing your results with others. It will be interesting to see if everyone made the same inferences about the unknown subject. These questions could help start your discussion.

a. How can you tell the difference between an observation and an inference?
b. Did any two people make the same observation but get a different inference from it?
c. Are observations or inferences easier to make?
d. How can you become more certain that an inference is probably correct?
e. What use are observations? What use are inferences?
f. What inferences have you made outside the classroom in the last day?

Problem 2-3

How much can you learn by observing living things?

In Problems 2-1 and 2-2, you made some observations and inferences about people. You could spend the rest of your life doing this. If you concentrated on human behavior, you might become a **psychologist** (sy-KOL-uh-jist), or behavioral scientist. Or you could limit your observations and inferences to other characteristics of people, like the reactions that go on inside of them. In this case, you might become a **physiologist** (fiz-ee-OL-uh-jist).

If you learned to be really skillful in planning and making observations and inferences, you could get paid as a professional scientist. If you became a **botanist** (BOT-uh-nist), you would make observations and inferences about plants. If you became a **geologist** (jee-OL-uh-jist), you would make

FIGURE 2.6
What can you infer from these photographs? Is one rabbit older than the other? Are these rabbits wild or tame? Are they related to each other?

FIGURE 2.7
What can you infer about these plants? Were they grown in the same soil? Were they grown from the same batch of seeds?

FIGURE 2.8
What can you infer about these animals?

observations and inferences about the earth. In fact, the main difference between one scientist and another is the subject each chooses to observe and make inferences about.

Why does a professional scientist make inferences? For the same reasons you do—to explain things that are puzzling and to show how different things or events may be related.

In this Problem, you will make observations and inferences about a plant and an animal other than a human. You will have a chance to improve your skill at making observations and inferences.

Materials
plant and animal
meter stick
magnifying glass
balance

Gathering data

Select either the plant or the animal. In 15 minutes, make as many observations and inferences as you can about the organism. You may find that rulers, balances, and magnifying glasses are useful in making observations.

Now, switch your attention to the other organism. Take another 15 minutes to make observations and inferences about this organism.

Recording data

Record observations and inferences in separate columns as you did in Problem 2-2. Plan on using different pages for the animal and the plant.

Analyzing data

After making your observations and inferences, compare them with your classmates' results. You might consider these questions when you discuss your results.

a. Are all the observations really observations? Or did you mix inferences in with them?
b. Did you make more than one inference from some observations?
c. A **quantitative** (KWAN-tih-tay-tiv) observation reports a quantity or measured value. How many of your observations are quantitative?
d. An observation that does not include a measurement is called a **qualitative** (KWAH-lih-tay-tiv) observation. It reports qualities or characteristics of something. How many of your observations are qualitative? Which observations are less likely to be misunderstood—quantitative ones or qualitative ones?

Problem 2-4

How fast does your heart beat?

In studying animals, scientists often observe heartbeat. Medical doctors listen to a patient's heartbeat as a general check on his health. That is why scientists measure the heartbeats of astronauts in space. Other scientists study the effects of new drugs on the heartbeats of laboratory animals.

Remember the imaginary scientist from Mars in Problem 2-1? If he observed Earthlings from a distance, he would probably not notice a human heartbeat. However, if the Martian scientist got very close to an Earth subject, he might notice the steady thump, thump, especially if Martians don't have heartbeats. But the Martian scientist would probably notice the heartbeat anyway. Part of being a good scientist is to observe even common things that hardly seem worth mentioning. In fact, heartbeat may have been one of your observations (or inferences) in Problem 2-1 or 2-2.

We don't know how a Martian would describe heartbeat. How would you describe your own heartbeat at the moment? You might say, "It's beating kind of slow but regular." That would give some information, but not very much.

You might say, "It's beating faster than when I'm falling asleep, but slower than when I'm running." That would give more information than the first try. However, it still isn't very exact.

By now, you have probably thought of a better way to describe your heartbeat. Make the description quantitative: Measure it.

In this Problem, you will observe and record your heartbeat quantitatively. You can decide yourself what methods to use.

The purpose here is not to get involved in a deep study of the human heartbeat. The main purpose is for you to improve your skill in making scientific observations.

Materials

Ask your teacher for any materials you think you need.

Gathering data

Choose a partner. Decide on a method to observe his or her heartbeat quantitatively. Remember that "observing" does not always mean just "seeing." Any or all of your senses can be used in making observations. You may want to make a simple **stethoscope** (STETH-uh-scope) by rolling a sheet of paper into a tube. This simple instrument would help you to hear heartbeats.

Will your observations involve measurements of time? If so, you will probably want to use some kind of clock.

Whatever method you use, you should repeat it several times. Record each observation. Try to keep all conditions

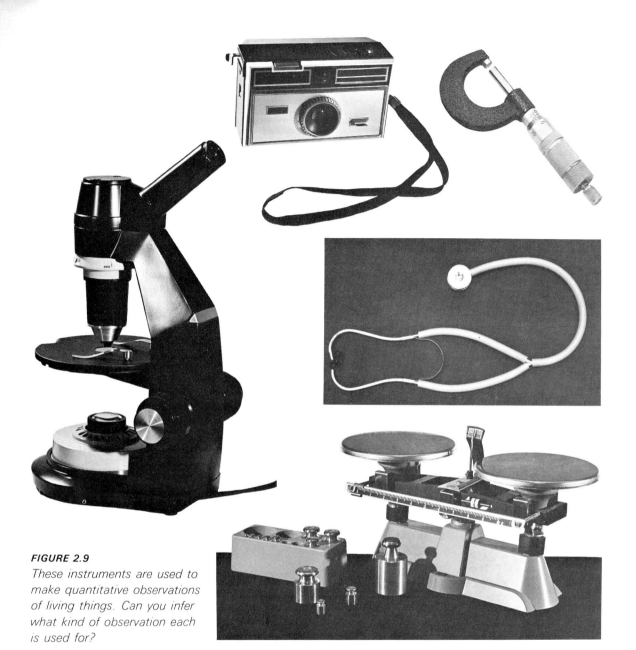

FIGURE 2.9
These instruments are used to make quantitative observations of living things. Can you infer what kind of observation each is used for?

the same from one observation to the next. Things such as room temperature, noises, and your own movement might change the heartbeat. You can probably think of other conditions that should be kept the same. After you have observed your partner's heartbeat, he should observe yours.

Recording data

You should make some kind of table to organize your data. The exact form is not important, but the table should be

FIGURE 2.10

The complex instruments on both sides of the patient makes very accurate observations of heartbeat. Observations are automatically recorded on the moving paper strip (right corner). How is this method superior to the usual method of observing heartbeat?

easy to understand. Figure 2.12 shows two ways you might arrange your data. Notice that the observations are called "trials" and are numbered.

Plan to make only one data sheet. Why isn't it a good idea to copy data from one page to another?

Remember! Record *all* data. Describe your method for observing heartbeat. If you use more than one method, be sure to note that.

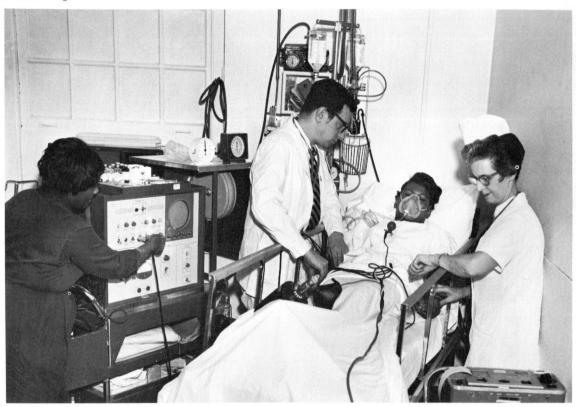

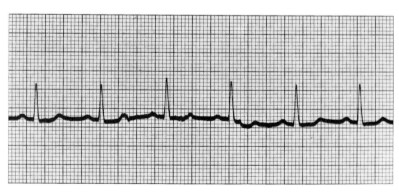

FIGURE 2.11

A permanent record of observations of the human heartbeat. How many beats do you think were recorded?

Making scientific observations

Analyzing data

Scientific observations by themselves aren't of much value. It's what you do with the observations that really matters. After you have made your observations, discuss the results with your classmates and teacher. You might use these questions in your discussion.

a. Did everyone in your class report the same heartbeat? Is this what you would have predicted?
b. Do some methods of observing heartbeat seem better than others?
c. Can you compare observations made by different methods?
d. Do all observations made by the same method on the same subject give exactly the same result? If not, what would be the best observation to report?
e. How do boys' heartbeats compare to girls' heartbeats?

FIGURE 2.12

Use a table similar to either of these, or think of another kind.

Data table – Heartbeat
Method 1 – count pulses in throat for 15 seconds
Subject: G.M.

Trial	Pulses in 15 sec.
1	17
2	19
3	15
—	—
—	—

Method 2 – listen to heartbeat with stethoscope and count beats for 30 seconds
Subject: G.M.

Trial	Beats in 30 sec.
1	35
2	33
—	—
—	—

Problem 2-5

How often do people blink?

Scientists don't understand completely why people blink. They know that a puff of air can cause a blink and so can a flash of light. The eye also blinks during normal activities, such as reading. In this case, blinking helps to keep the surface of the eye wet. Some scientists claim that blinking is sometimes caused by nervousness.

In this Problem, you will observe blinking. Your observations should be quantitative. Also, you should try to find the typical or usual rate of blinking for the people you study. This is different from your observations of heartbeat. Then, you studied only one person. You couldn't make a general statement about heartbeat from the data on one subject.

FIGURE 2.13
Does this girl really talk with her eyes closed? Her friends probably didn't even notice that she was blinking. However, the camera caught it. How can a camera be useful in making scientific observations?

Making scientific observations 27

Blinking can be observed easily. Therefore, it may be useful in studying humans and certain other animals. You may become interested in the rate of blinking and its causes. However, the main reason for studying it here is to develop your skills as a scientific observer.

Materials
clock

FIGURE 2.14
A wink or a blink? Which do you infer? What reasons do you have for your choice?

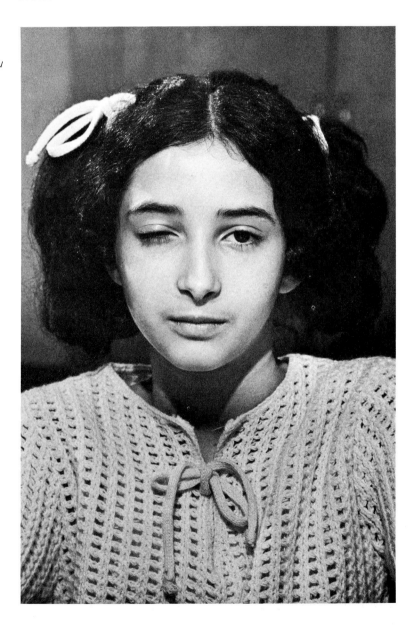

Gathering data

Decide on a method for observing a classmate's blinking. Everyone can probably control his blinking to some extent. Therefore, you will need to think of a way to help your subject relax. Then you can get a normal count. For instance, you could give the subject a simple task to do like reading or doing some arithmetic problems. While he is busy, you can observe him blink.

Your method should not take a long time. You should test several people in the time available. Also, part of the time you will probably be another classmate's subject.

Before you start, you should think about your overall plan. How many subjects will you observe? Will all the subjects be boys? How long will you observe each subject?

Recording data

Record your data on one page. It will help if you design a table for the data *before* you start to observe. You may want to make one column for quantitative observations and another column for unexpected observations. In this second column you would note things like "subject sneezed twice," or "subject started laughing."

Analyzing data

After you get your data, take a few minutes to study it. Can you now answer the question, "How often do people blink?"

Write your answer on the bottom of your data sheet. Also, explain briefly how you figured out your answer. Or you might want to explain why you can't give a final answer to the question.

Look at the data for the whole class. What can you and your classmates conclude about blinking? The following questions might help you compare your results with your classmates.

a. What were the fastest and slowest rates of blinking?
b. Do boys blink more often than girls?
c. Would you expect to get the same results if the experiment were repeated with another class or with adults as subjects?
d. What method seemed to produce the best observations?

Making scientific observations

Mastery Item 2-1

Describing an animal

Your teacher will show you an animal. In five minutes, make as many observations of it as you can. Write each observation briefly on a sheet of paper.

Key

You have mastered this Item if you made 10 or more observations, and if no more than one of your "observations" is really an inference.

Mastery Item 2-2

Interpreting a photograph

Observe Figure 2.15. Make as many inferences as you can about the photograph. You can take 10 minutes. Divide a sheet of paper into two columns. Write your observations in one column and your inferences in the other column.

FIGURE 2.15
What inferences can you make about this photograph?

Key

[printed upside down] You have mastered this Item if you make eight inferences in the time allowed. You and your teacher together should decide whether the inferences are reasonable.

Mastery Item 2-3

The truth about two trees

Look at Figure 2.16. The following list contains a mixture of observations and inferences about it. Decide which statements are observations and which are inferences. Mark each of the statements either **O** for observation or **I** for inference.

a. The trees are the same height.
b. The trees are the same age.
c. The picture was taken in the summer.
d. The color of the leaves is green.
e. One of the trees is sick.
f. One tree has more leaves than the other.
g. Both trees are growing in the same kind of soil.
h. The sun was shining when the photographer took the picture.
i. The grass has been mowed regularly.
j. The trees are the same kind.

Key

[printed upside down] Remember that an inference (I) goes beyond a direct observation (O). One might reasonably answer the statements as follows:

a.—O, b.—I, c.—I, d.—I, e.—I, f.—O, g.—I, h.—O (or I, The picture could have been taken using a strong flash unit), i.—I, j.—I.

If your answers do not agree with these, discuss them with your teacher. Other answers could be correct, depending on your reasons.

Making scientific observations

FIGURE 2.16
What statements could you make about these trees?

Mastery Item 2-4

An inference about fish

A person went to a pet store to look at tropical fish. He became interested in one tank in particular. The label on the tank said "brick swordtails." The person counted the number of fish in the top half of the tank and the number of fish in the bottom half of the tank. He waited a minute and then counted the fish again in each half of the tank. The observer repeated this procedure several more times. Figure 2.17 is a record of his observations.

Make one good inference from the data. You and your teacher together should decide whether your inference is a good one.

Key

One possible inference is that these fish prefer the bottom of the tank. Or, the fish "like" colder water or greater pressure. There might be plants growing in the bottom of the tank. You could infer the fish like to hide among the plants.

The statement, "There are 12 fish in the tank," is definitely an observation, not an inference.

FIGURE 2.17

Can you make an inference from this data?

Number of tropical fish in the tank. Trials one minute apart.

Trials	Number in upper half	Number in lower half
1	3	9
2	0	12
3	3	9
4	3	9
5	3	9
6	2	10

Mastery Item 2-5

How fast do people breathe?

Imagine that you are going to study human breathing. You have decided that you need to observe exactly how fast people breathe. Describe the method you would use to make this observation quantitatively. The method you describe should give accurate results.

Key

You should have described most of the following things:

a. Special equipment, if any would be needed.
b. The length of time for observations and the number of observations.
c. The number of people you would observe, and what they would be doing while you measured their breathing rates.

If you and your teacher agree that your method is good, you have mastered this item.

FIGURE 3.1
Look at this drawing for a few minutes. Does anything happen?

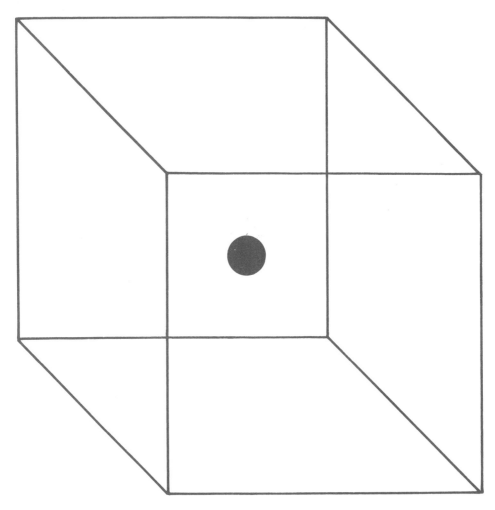

Investigation

3 Perceiving the world

In Investigation 2 you learned how to make observations and how to tell observations from inferences. To infer is to explain or interpret the facts. Good observations, on the other hand, are "the truth, the whole truth, and nothing but the truth." Or are they? Look at Figure 3.1 for two or three minutes.

Since the figure does not change, how can you explain why your observation changes?

Actually this is a complicated question with no firm answer. Many factors affect observations, and that is what Investigation 3 is all about.

Perhaps your eyes are playing tricks on you. Or maybe the mind plays tricks with what the eye sees! Your observation of the Necker cube must be affected by *something you do* when you look at it. The lines on the paper don't move. At least sometimes, therefore, making observations is complicated.

There is a special word to describe such observations. The word is **perception** (per-SEP-shun). To have a perception (to perceive) means to become aware of things through your senses. Is it also possible that things we smell, or hear, or feel aren't exactly what we perceive them to be?

By doing this Investigation, you will learn how some living things get information about their surroundings. You will also learn how the information is shaped into perceptions.

First read about four different kinds of animals and how they perceive their environment. Then you will see which factors affect *your* perceptions of your environment.

Problem 3-1

How do animals get information?

Gathering data

Read about some unusual perceptions in the following pages. You will then be asked to discuss what you have read.

BAT NAVIGATION IN THE DARK

Many kinds of animals can tell where they are by echoes. They make sounds, and then listen for the echoes from nearby objects. The laboratory rat can do this quite well. Bats and porpoises can also use sound in this way. "Echo location" is their usual way of finding food and avoiding obstacles.

A bat finds his way through the air by squeaking out a stream of high-pitched sounds and listening for the echoes. The returning sounds are interpreted by a special area of the brain. The method is like radar. A bat can tell the size, shape, and distance of objects around him even in the dark. If the object is an insect, the bat can fly toward it. If the object is a tree or a wire, the bat can fly around it.

Parts of the bat's body allow it to use echo location. A bat has a huge voice box that is good for making high-pitched sounds. Bats also have very large ears that can detect the sounds made by flying moths.

Imagine, for a moment, you could hear like a bat! Imagine being able to hear a moth! If you could hear that well, how would your perception of your environment change?

FIGURE 3.2
Notice the very large ears on the bat's head. A bat can hear the sound of a moth flying.

SNAKES THAT HUNT BY HEAT

Did you ever hold your hand close to a stove or an iron to find out if it was hot? What you were doing was perceiving the invisible heat rays it was giving off. You have sense organs or **receptors** (reh-SEP-ters) in your skin that detect these rays.

There is a large group of snakes that can locate warm objects by the invisible heat rays these objects give off. This group of snakes includes rattlesnakes and the blind boa, which lives in caves. In front of their eyes these snakes have special pits or holes, which pick up the rays from warm-blooded animals. Warm-blooded animals, like people, mice, or birds, have a steady, high body temperature.

Figure 3.3 shows the two heat-sensitive pits in the head of a rattlesnake. Notice how the location of the pits allows rays to come in only from certain directions. These receptors can spot a warm object in cool surroundings from two feet away.

How could a rattlesnake use his heat-sensitive pits? Do you think you would feel differently about the room you are in right now, if you had the snake's special pits?

FIGURE 3.3
The rattlesnake has two pits in its head. The pits can detect heat given off by nearby animals.

FIGURE 3.4

A female moth in a cloth cage attracts many male moths.

HOW DO MOTHS COMMUNICATE?

A French biologist, **Jean Henri Fabre** (zhahn ahn-REE FAH-br), had a female moth that he wished to study. He placed the moth in a cloth cage near an open window in his house. Within 15 hours, 60 males of the same kind of moth collected around the cage!

Fabre wanted to find out how the male moths located the female. He placed the female moth near the same window, but this time in a tightly closed glass container. Now she could only be seen by other moths. No male moths appeared.

Fabre repeated these experiments with another kind of female moth and obtained the same results. When the female was in a cloth cage, she attracted males from as far away as a mile or two. But when she was placed in a clear glass container, no male moths appeared. Why did this happen?

Moths have receptors called **antennae** (an-TEN-ee) on their heads. Fabre noticed that males with their antennae removed could not locate the female in either cage. He then reasoned that perhaps the female moths send out vibrations. These vibrations might be picked up by the antennae of the male moths, and the males could then be attracted. Males without antennae, therefore, would not be able to locate females because they could not detect their vibrations.

FIGURE 3.5
No male moths gather around a female moth which is placed in a glass jar.

One piece of evidence puzzled him, however. Male moths also flocked to containers from which he had recently removed a female. Maybe the female had an attractive odor! But Fabre couldn't see how the males could smell a female moth from several miles away.

How do you think male moths find female moths? Can you think of a way to find out whether the males are attracted by an odor or by a vibration?

Did Jean Henri Fabre have the same perception of the environment as the moths?

AN ELECTRIC PERCEPTION SYSTEM IN A FISH

Gymnarchus nilocticus (jim-NAR-kus ny-LOK-ti-kus) is a fish that lives in the muddy water of African rivers. *Gymnarchus* is an active fish, and it feeds mostly on smaller fish. Studies indicate that it feeds at night and hides under logs or rocks during the day.

Close observations of *Gymnarchus* in a darkened aquarium show that its movements are well controlled. When it is feeding, it never bumps into obstacles or the walls of the tank. It moves easily forward or backward in spite of the fact it can see only in bright light. How can *Gymnarchus* be so efficient when visibility is low?

Biologists have known for some time that this fish has organs that produce weak electric currents. But the function of these electric organs was not understood. A biologist investigated the function of these organs. He placed *Gymnarchus* into a tank that contained electric measuring equipment. He discovered that *Gymnarchus* surrounds itself with a weak electric field. We think this electric field looks something like Figure 3.6. You could create a field that looks similar using a bar magnet and some iron filings.

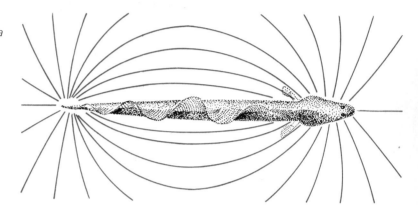

FIGURE 3.6
Gymnarchus *is surrounded by a weak electric field. Scientists think it may look like this drawing.*

FIGURE 3.7
Compare the Gymnarchus, *above, with the sunfish at right. How are they different from each other?*

FIGURE 3.8

Changes in the electric field of a Gymnarchus. *Lines of current from the* Gymnarchus *flow around a stone in* a. *In* b, *the lines of current flow toward and through a fish. Thus the fish can tell the difference between living and nonliving things.*

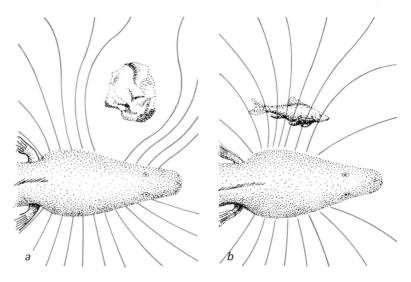

Figure 3.7 shows *Gymnarchus* and a fish that you may recognize, a sunfish. As you can see, *Gymnarchus* does not have the broad tail that most fish use to push themselves through the water. Instead it moves either backward or forward by waving its long fin. In that way *Gymnarchus* can always keep its body straight while moving, and not disturb its electric field.

Gymnarchus locates nearby objects when they bend this electric field (Figure 3.8). Special receptors in the head of the fish measure any changes that take place in the field. *Gymnarchus* can also tell the difference between living and nonliving things. They change the electric field in different ways (Figure 3.8).

What advantages does an electric perception system have for this fish? If *Gymnarchus* has a picture of his environment in his brain, can you imagine what it is like?

Analysis of data

Discuss these questions with other members of your class.

a. What kinds of information do animals get about their environment?
b. What determines the information an animal can receive?
c. Why is it hard to imagine what a fish, bat, or moth perceives?
d. What factors other than sense receptors might influence perception?

Problem 3-2

How reliable are your senses?

In Problem 3-1 you learned that different kinds of animals receive different information from their environment. A grasshopper, a frog, and a person may perceive the same meadow quite differently. Now let's consider another problem about perception: How good is the information that we receive? Do our sense organs tell us exactly what our environment is like? Do we all get the same information from our environment? The activities in Problem 3-2 will supply some answers to these questions. Your group may do some or all of the activities, depending on the time and supplies that are available.

Does everyone have the same ability to taste?

Some people in your class laugh louder than others. Some people have longer arms than others. We know that no two people are alike. Do we also have different perceptions of the same things?

Try to find out if everyone has the same tastes.

Materials
Your teacher is going to give you a substance to taste and directions for tasting it.

Gathering data
While everyone in the class is tasting the substance, look around you at the reactions of your classmates.

Recording data
Write down your reactions and those of your classmates.

Analyzing data
Does everyone in your class have the same ability to taste? Why do you like the foods you do? Is it because you grow to like them or because you have a special ability to taste?

Perceiving the world

FIGURE 3.9
Outlines for parts **A**, **B**, and **C**, to be traced and cut out during the next exercise.

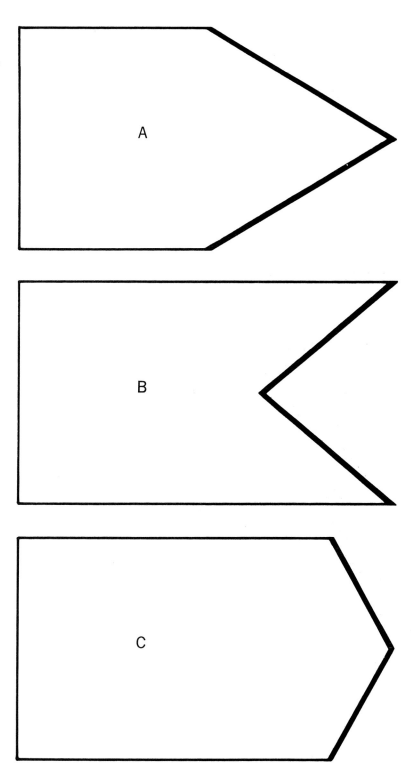

How accurately can you see?

Are you good at judging distances? You can perform the following exercise and find out.

Materials
cardboard or heavy paper
scissors, paste, and pencil
felt-tip marking pen or crayon
tracing paper
unlined paper
ruler

Gathering data
You will be working in small groups during this exercise. Each group will prepare the necessary materials. Trace the pieces in Figure 3.9 and cut out your tracings. Use these cutouts as patterns and make two of each of the pieces out of heavy paper or light cardboard.

Next follow the directions for Figure 3.10. Mark heavy lines on the playing sheet with a crayon. Then lay the sheet down on a table or desk.

To begin the experiment place the **A** pieces on the top line, any distance apart, as shown in Figure 3.11. Then place the **B** pieces on the bottom line. Move the **B**'s until you think they are exactly as far apart as the two **A**'s. Now have another member of your group measure and record the lengths of line **A-A** and line **B-B.**

Repeat the procedure two more times with a different distance between the **A**'s each time.

When the whole group has completed these measurements, substitute the **C**'s for the **B**'s. Now test your ability to estimate the length of the lines with the **C**'s. Carefully record three sets of measurements for each group member.

Recording data
Make your own group chart for recording data. Your chart should have four columns. Record distances for each **A-A** line and **B-B** line that each person in your group matches. Do the same for each **A-A** and **C-C** line matched. Record the measurements for each trial on your chart.

Analyzing data
Look at your data. How well do the **A-A** and the **B-B** distances match? How well do the **A-A** and **C-C** distances

Perceiving the world

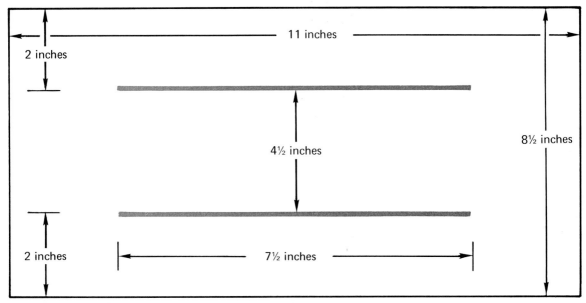

FIGURE 3.10
Make two thick 7½ inch lines on a sheet of 8½ × 11 inch plain white paper. These lines should be 2 inches from the edge of the paper.

FIGURE 3.11
Place the cutout pieces on the lines you drew, as shown here.

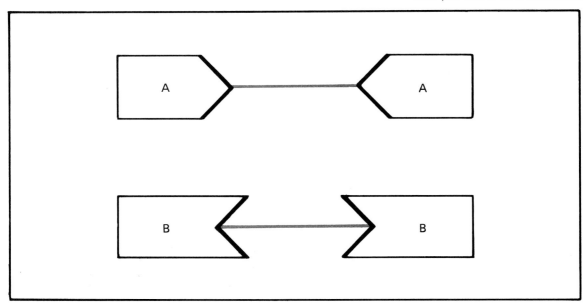

match? Does the accuracy of your estimates seem to depend on the shape of the pieces? If so, why do you think this happens? In your own words write out the best explanation of the data.

Discuss your explanation with other members of your class and your teacher.

Are the estimates made by the girls in your class more accurate than those made by the boys? How could you organize your data to find out? Can you find out the average error of your class when matching line **A-A** with **B-B**?

Is the water hot or cold?

When we say *the weather* is cold or *the weather* is hot, what do we actually mean?

We all have receptors for heat and cold located in our skin. How accurate are these receptors? Do they give us reliable information?

Materials

Your teacher has set up some containers of water.
thermometers

Gathering data

Your teacher will tell you how many students will perform this activity. If you are chosen, first place one hand in pan **H** and one hand in pan **C**. (See Figure 3.12.) Then, after three minutes, place both hands in pan **?**. Tell the class what you feel. Is the water hot or cold? Is the data you get from each hand the same? Now measure the temperature of the water in each pan with a thermometer.

The thermometer you use may measure temperatures in the Celsius system (°C) or the Fahrenheit system (°F), or even both. The United States hasn't agreed to use one system all the time. Scientists nearly always use the Celsius system to record data. Other people generally, but not always, use the Fahrenheit system.

Recording data

Write your responses in your notebook and summarize your classmates' responses. Also, record the measured temperatures of the pans of water.

FIGURE 3.12
How can you test to see if water is hot or cold?

Analyzing data

Suppose you had to describe to another person the water temperature of pan **?**. Which data would you use and why?

Determining the direction of sound

You just felt the left hand and the right hand disagree about water temperature. Suppose you set up a similar experiment on hearing. Your ears might disagree on whether a sound was loud or soft. Hot, cold, loud, soft—these are matters of personal judgment.

Would you expect your ears to accurately tell you the direction a sound came from? That, after all, would not be simply a matter of personal judgment. You could check your perceptions.

Materials

clickers or other noisemakers   blindfold
chairs   chalk

Gathering data

This experiment requires groups of nine students. One person, the test student, sits in the center of a circle of eight other persons. Those on the outside are numbered, one to eight. (See Figure 3.13.) Each has a noisemaker. The test student sits blindfolded, facing number one. When everyone is seated as in Figure 3.13, you are ready to begin.

Student number one points out another student to make a sound. This student clicks his noisemaker, and the test student tries to point in the direction the sound came from. The test subject should not turn his head. Sounds should be made about every fifteen seconds. If you are student one, do not select noisemakers in order. Why? You should take

FIGURE 3.13
Can the blindfolded person in the middle tell which student is making a sound?

Perceiving the world

turns yourself, too. Check the test student on 15 sounds. Student number five records the accuracy of the guesses. (See Figure 3.14.)

Test as many students as time allows—at least four or five.

Recording data

Record each person's perceptions on a chart like Figure 3.14. Summarize class data in a table like the one in Figure 3.15.

Analyzing data

Are the sounds from some places on the circle predicted more accurately than others? Can you develop an explanation to account for the differences?

FIGURE 3.14
You can record your test data this way. Each time the test student guesses correctly, place a **+** next to the proper number on the chart. If he guesses incorrectly, place a **−** next to the proper number.

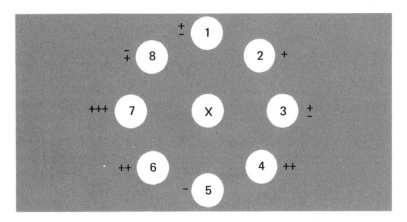

FIGURE 3.15
To organize your data, count the number of **+**'s and **−**'s for each position, and place the numbers in the proper columns of a table.

Position number	Correct response	Incorrect response
1		
2		
3		
4		
5		
6		
7		

Problem 3-3

How does experience affect perception?

So far in this Investigation you have looked at the kinds of information animals get about their environment through their sense receptors. This information is only the raw material for perceptions. For example, your eyes only notice black marks on paper, but you perceive them as words. Thus, more than sense organs are involved in perception. In the same way, you can tell the sound made by a dog from one made by a bird.

Does what has happened to us in the past affect the way we perceive the present? The next three activities will help you answer this question.

What don't you hear?

Can you recall all of the objects your eyes saw on the way to school this morning? Or all of the sounds you heard? What factors affect what we see and hear?

Gathering data

Do your ears pick up more noises than you are aware of? What factors affect what you hear and what you do not? Discuss these questions with your classmates. During the discussion the teacher will turn on a tape recorder.

Afterward, the teacher will stop the recorder and replay the tape. Listen for sounds in the room you weren't aware of during the discussion. How many sounds were there?

Analyzing data

Do you think you often select what not to hear? How do you decide what to listen to?

If you are interested in problems about sound and hearing, you might investigate questions like these: Does your hearing affect your skill at ping-pong? What is the noisiest room in your house? Does noise affect your ability to concentrate?

Perceiving the world

FIGURE 3.16
How can these two students concentrate on their studying when there is so much noise in the background?

How do we perceive sizes?

Did you ever wonder why the tiny people on television do not seem to be tiny? Or, why is it when you look down on a car or house from a tall building they seem so small? If you do this Problem, you might be able to answer these questions.

Materials
tracing paper
cardboard or heavy paper

FIGURE 3.17
This photograph of a straight road was taken in Australia. What happens when you place cutouts of the car in different places on the road?

Gathering data

Figure 3.17 shows a road running into a desert. Next to the Figure is the outline of a car. Trace this outline on thin paper, then cut out your tracing. Use it as a pattern and cut out of heavy paper two such cars. Place one of the cars over the letter "A" on the photograph. Place the other car over the letter "B". Which car seems bigger? Move the cars around on the picture. Do you always get the same effect? What clues help you perceive the size of objects?

Now look at Figures 3.18 and 3.19 and see if you can learn more about size perception.

Recording data

Does the building with the arch through it in Figure 3.18 seem to be close or far away? Write down all the clues you can find in Figure 3.18 which help you determine its distance.

Is the person in Figure 3.19 an adult or a child? Give any data you can find to support your statement.

Analyzing data

Summarize the clues you used to determine the size of objects. To refresh your memory, briefly look over the exercises you just did on determining sizes.

FIGURE 3.18
How tall is the archway in this picture? Can a tall man fit under it? A large truck?

FIGURE 3.19
Can you tell how old this person is, or whether it's male or female? What clues do you have?

Does everyone perceive in the same way?

When many people are shown the same object, do they all perceive it the same way? Does a photograph or picture have the same meaning for each viewer?

Materials
Pictures will be projected on a screen.

Gathering data
In this Problem you will be shown four pictures. Each picture will be projected for 10 seconds. Write down what you think you see. The teacher will then show you each picture again. This time you can study it as long as you like and compare your interpretation with your classmates'.

Recording data
One student will record on the blackboard what the class saw in each picture. Try to make your description as brief as possible.

Analyzing data
Did everyone see the pictures the same way? Would everyone see a person, a classroom, a street, or a river the same way?

Is disagreeing about the meaning of a picture the same as disagreeing about the length of a line or the temperature of a room? What kinds of things should people agree on?

You may be coming to the conclusion that most people develop their own personal perceptions of the world and that these perceptions may often be quite different. If you are thinking this way, the Investigation has been a success!

Mastery Item 3-1

Perceiving a city street

The paragraph below describes the behavior of a 52-year-old man who regained his sight after being blind.

"Before getting back his sight, S. B. was unafraid of traffic. He would cross the street alone, holding his arm or cane *stubbornly before him, until the traffic would stop. . . . But after the operation, it took two of us, one on either side, to force him across a road. He was terrified as never before in his life."*

(Modified from Gregory, R. L. *The Eye and the Brain.* New York, McGraw-Hill Publishing Co., 1966.)

Why do you think this man's perception of the street changed?

Key

Apparently, S. B.'s perception of the street changed when he was able to see. He had more information now about traffic than when he was blind. His new visual perceptions were perhaps too strange and frightening to allow him to act confidently.

Mastery Item 3-2

Which way is up?

Look carefully at Figure 3.20. Write a description of what you see. Now turn your book upside down. Again look carefully at Figure 3.20 and write a description of what you see.

FIGURE 3.20
What do you see in this photograph? What happens when you turn the book upside down?

Now compare your two descriptions of the photograph. If you were a scientist working for the space administration, how would you interpret this photograph?

Key

You were correct if you said that at least two interpretations are possible. Viewed with the book in a normal position, the moon's surface seems to have many raised areas or plateaus. Upside down, the plateaus become craters or holes.

Why do you have two perceptions of the same photograph? Is it because you have learned that objects usually are lighted from above? With overhead lighting, a bump is brightest on top and darkest at the bottom. A hole, lighted the same way, has a shaded area on top and is brightest at the bottom.

To interpret these photographs accurately you would need to know the direction of the sun at the time the picture was taken.

Mastery Item 3-3

A perception experiment

A scientist concerned with perception designed several experiments to gather data. One of his experiments is shown in Figure 3.21.

FIGURE 3.21
Can you think of any way to explain what you see in this photograph?

Perceiving the world　　　　　　　　　　　　　　　　　　　　**57**

Your task in this Mastery Item is to give the best and most logical explanation you can for your perception of Figure 3.21. Write your answer as carefully and as accurately as possible.

Key

Is the person at the right really twice as large as the person on the left? It would be difficult to say exactly, using only this photograph. But you could have gotten some clues if you measured parts of the room. If you did, you were successful in this Mastery Item. For you found out by measuring that the room is not a normal room. The right wall is longer than the left wall. Also, the window on the right is larger than the window on the left. Why does a room like this affect your perception? Do you think any other part of the room is not normal?

Mastery Item 3-4

A real flying saucer?

Look at the photograph taken by a young man with a simple box camera (Figure 3.22). What do you think it could be? On the basis of the evidence you can get from the photograph, write out your perception of the object. How might other people perceive it?

Key

The data available in Figure 3.22 does not allow an accurate description of the object. You were correct if you said many perceptions were possible. It could be a photograph of a plastic plate that was thrown into the air. It could be a flying saucer. It could be the sun reflected from a cloud. Perhaps the best answer might be an Unidentified Flying Object.

FIGURE 3.22

This photograph was taken with a simple box camera. What do you think it is?

FIGURE 4.1
What classification system is used here?

1968 GALAXIE "500"
4-door hardtop with 8-cylinder engine, Cruise-O-Matic transmission, radio, and power steering **$2388**

1968 MUSTANG
Equipped with 8-cylinder engine, 4-speed transmission, radio, etc. Beautiful red finish. FACTORY AIR CONDITIONING **$2288**

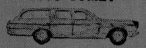
1968 COMET
Villager Squire. Equipped with V-8 engine, radio, Merco-Matic transmission, power steering, power brakes. Beautiful dark green finish. Only **$2088**

1966 M.G.B. GT
Equipped with radio, heater, 4-speed transmission, wire wheels. EXTRA SHARP. ONLY **$1888**

1967 CHEVROLET
Impala, V-8 engine, Powerglide transmission, radio, power steering, FACTORY AIR CONDITIONING. ONLY **$1988**

1962 CADILLAC
B628. Coupe DeVille, radio, hydramatic, power steering, power brakes, beautiful yellow finish, EXCEPTIONALLY NICE. **$988**

1965 MUSTANG
Hardtop with 6-cylinder engine and standard transmission **$988**

1966 MUSTANG
Hardtop with burgundy finish, 8-cylinder engine, 4-speed transmission and radio **$1488**

1967 MUSTANG
Hardtop with automatic transmission, radio and Lime Gold finish **$1788**

1966 FORD
Ranch Wagon with Cruise-O-Matic transmission, 8-cylinder engine, radio and power steering **$1488**

1966 FORD
Country Squire Station Wagon with 8-cylinder engine, Cruise-O-Matic transmission, power steering and radio. FACTORY AIR CONDITIONING. Beautiful red finish with roof rack **$1988**

1967 FORD GALAXIE "500"
HARDTOPS—CHOICE OF 3
V-8 engines, radio, Cruise-O-Matic transmission, power steering. Choice of 3 different colors **$1888**

1964 FAIRLANE
2-DR.—CHOICE OF 2
Cruise-O-Matic transmission. ONLY **$788**

1967 RAMBLER
Ambassador 2-door hardtop with radio automatic transmission and power steering **$1888**

1966 FORD LTD
4-door hardtop with black vinyl roof, 8-cylinder engine, radio, Cruise-O-Matic transmission and power steering. Choice of 2 .... **$1788**

1965 MERCURY
Beautiful Convertible with Merc-O-Matic transmission, radio and power steering. Sharp white finish **$1288**

1967 COUGAR
V-8 engine, merco-matic transmission. Beautiful red finish offset by black vinyl roof **$1988**

1966 ECONOLINE CLUB WAGON
Radio, Cruise-O-matic, Fords most versatile unit that carries people, stuff and things .. **$1488**

Investigation 4

Organizing information

The process of classification is very useful. You probably use it all the time without giving it a name. For instance, you may classify television shows into groups such as "good" and "bad." Or maybe you use a three-part system, such as "excellent," "fair," and "terrible." Probably you also classify them by type, such as "western," "science fiction," "sports," or "musical." As you can imagine, the number of ways of classifying TV shows is almost endless.

In science, classification is used a lot. Scientists classify things for the same reason you classify TV shows. It makes thinking about a large number of things simpler and easier.

There are other ways you have used classification without calling it by name. Do you collect stamps or coins? Stamp albums help you group or classify your stamps by country. Coin folders help you to group your coins by their face value, that is, by whether they are pennies, nickels, or dimes. If you collect shells or rocks, you may use a classification system of your own based on color, size, or where you found them.

Think of many ways you have been classified by other people. In school, you have been classified as a first grader, second grader, and so on. Much of your school work has been classified by the teacher as "A," "B," or "C." Older students are classified in many ways also. You have probably heard and used terms like "hippy," "brain," and "athlete." Each of these terms represents somebody's classification system for students.

In this Investigation, you will learn and practice classifying as it is used in science. Then you may be able to recognize classification systems when you come across them in everyday life. You will also learn how to use classifications to make reasonable guesses about how and why animals do certain things.

Problem 4-1

How can things be classified?

In this Problem you can practice your skills of grouping or classifying. You will mainly be classifying collections of living things. However, the process of classifying can be applied to nonliving things equally well.

Materials
several collections of various living, once-living, and non-living things

Gathering data
Carefully observe the collection that your teacher shows you. Divide the collection into two groups. Each group should have some characteristic that the other doesn't. There should be at least two individuals in each group.

Once you have formed two groups, try to subdivide each of them into two more groups, using another characteristic. As before, you should have a clear reason for your grouping.

Now, put all the specimens back into one group. Repeat the process of grouping and subgrouping. But this time try using different characteristics.

In the same way, classify as many other collections and items in the Figures as you have time for.

Recording data
For each collection you classify, describe the characteristic used to make the first grouping. Then list the individuals that you place in each group. Do the same thing for the subgroups that you make from each of the first two groups.

Organizing information

FIGURE 4.2
How many ways can you classify these stamps into two separate groups?

FIGURE 4.3
How could you classify the animals shown on these stamps?

Organizing information

m

n

o

p

q

r

s

t

u

v

w

64 Organizing information

FIGURE 4.4
How could you classify these plants?

Analyzing data

Compare your classifications with those of your classmates. Did you all use the same characteristics for making your groups? How would you answer these questions on classification?

a. How many possible ways are there to classify a given collection of organisms?
b. How many groups and subgroups can be made from a given collection? What's the smallest number that can be in a group?
c. Are there certain characteristics that can be used to group any collection of living things?

FIGURE 4.5
How could you classify these shapes?

Problem 4-2

Making hypotheses—why do different people have different heartbeat rates?

Classifying sometimes raises questions. But often it helps you find answers, too. For instance, imagine that you classified the students in your class into two groups depending on how long each person could hold his breath. Then you might wonder, "Why is a person in the 'long breath group' and not the 'short breath group'?"

One way to answer this question would be to look at the two groups again. You could try to find another characteristic that separates the groups. You might find this: Mostly tall people were in the "long breath group," and mostly short people were in the "short breath group." Maybe tall

FIGURE 4.6

How could you classify the animals described in the table? A hint to get you started: Arrange the masses in a list from smallest to greatest. Find the middle of the list. The animals on one side of the midpoint are "animals with smaller masses." Those on the other side of the mid point are "animals with greater masses." The same technique will work for speeds, ages, or other characteristics. What other plans can you invent?

Animals	Characteristics				
	*breathing rate	top speed (meters per sec)	maximum age (yrs)	average mass (kg)	heart-beats (per min)
elephant	155	4.50	77	2740.00	40
tortoise	35	0.04	150	136.00	10
man	220	9.00	115	68.00	70
cat	710	10.00	21	2.30	200
horse	250	15.70	50	408.00	45
water snake	30	0.22	18	0.45	35
rabbit	750	18.00	12	0.68	250
mouse	3500	0.45	3	0.02	620
dog	580	15.70	20	9.10	100

*The amount of oxygen an animal uses in an hour, divided by the animal's mass (cubic millimeters per gram per hour)

people can hold their breath longer than short people. This "tall-short" classification might make sense of the "long breath-short breath" classification.

At this point, it will be a good idea for you to learn the meaning of a word you may not know, **hypothesis** (hy-POTH-uh-sis). Hypotheses will be used quite often in this Problem and later Investigations. A hypothesis is a reasonable guess, a possible explanation for something. For example, we just suggested this hypothesis: Tall people can hold their breath longer than short people. Can you make another hypothesis to explain some people's ability to hold their breath for a long time?

Let's take the question asked in this Problem: "Why do different people have different heartbeat rates?" You had some experience in Investigation 2 with heartbeat rates. You might have noticed that boys' and girls' heartbeat rates seemed to be different. You might **hypothesize** (hy-POTH-uh-syz), or make the hypothesis, that boys' hearts beat faster than girls'.

Is this hypothesis probably right? The best way to decide is to gather some evidence. Then see whether the evidence supports the hypothesis. This is called "testing the hypothesis." In your class you could test the hypothesis that boys' heartbeat rates are faster than girls'. What evidence would you need? You might actually get this evidence later.

As you may have guessed, the process described above happens very often in science: ask a question, make a hypothesis, test the hypothesis. You should not get the idea that these are *the* three steps in science. Scientific investigations don't always follow the pattern. However, the pattern does occur often enough that it will be worth your time to learn it. The pattern is summarized in Figure 4.7 on the next page.

Now you might say, "OK—I can see how these three steps might work out. But how do you make a hypothesis once you have a question?"

This, in itself, is a good question, and there is no one answer to it. You could produce a hypothesis just by making a wild guess. If that were all you did, you would waste a lot of time testing hypotheses that were doomed from the start.

One good way to make hypotheses about living things is to classify them. In the heartbeat example described before, two groups were made—boys and girls. The purpose was to find out if the sexes could cause differences in heartbeat rates. Other groups may lead to other hypotheses that explain the observed differences better than sex does.

The principal activity in this Problem is to make hypotheses, then to collect data to test them.

Materials

instruments to accurately measure heartbeat rates for each person in your class, for example:
clock or stopwatch   stethoscope
instruments to measure characteristics that will be related, for example:
meter stick   calipers   scale

FIGURE 4.7

Steps in solving problems. a. Ask a question. How long can a person hold his breath? *b. Make a hypothesis.* A person's height affects his ability to hold his breath. *c. Test the hypothesis.* Measure heights and breath-holding times for a large group of people.

FIGURE 4.8
To find a pulse at the wrist, move your three middle fingers in the area shown until you feel the beating of your pulse. Use a watch or clock with a second hand and count your pulse beats for 15 seconds.

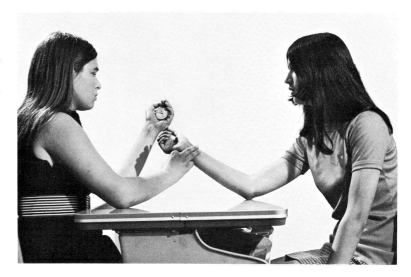

Gathering data

Think of a simple method to measure the heartbeat rate of all the members of your class. Figure 4.8 shows how to take a pulse. The exact method to use should be decided in a class discussion. Work in groups. Once each group gets data on its members, all the data from the class can be pooled. This way everyone in class will have the same data for every person's heartbeat.

Make a hypothesis to explain why some heartbeat rates are faster than others. Here's where your ability to notice clues could help. For instance, you might notice that the heaviest people seem to have the fastest heartbeat rates.

Of course, there are lots of other characteristics that you could use to group people. They might give you an idea for a hypothesis. Examples might be hair color or shoe size.

Now your imagination will come in handy. Can you think of a logical reason to connect the characteristic with heartbeat rate? Recall our sample hypothesis: Taller people have faster heartbeat rates. You might reason that the heart would have to beat faster to pump the blood higher in tall people.

Once you make your hypothesis, you will need to collect more data to test it. In our example, you need each person's height.

If time is available, make several hypotheses and collect the data necessary to test them. Use all the class time for making hypotheses and collecting data. Don't take the time to analyze your data now. There will be time for that later.

Recording data

Write down each hypothesis that you test. It should be a complete sentence. For example: Tall people have faster heartbeat rates than short people. Do this before you start gathering data to test the hypothesis.

Make a table for recording heartbeat data. Leave room to record the other data that you need, such as sex or height. Everyone in the class will have the same data on heartbeat rates. But you may record other data that no one else has.

Analyzing data

This part of this Investigation could be the most difficult because people often disagree on how to interpret the same data. This is also a very interesting part of the Investigation, maybe the most interesting. Here you will decide whether to accept or reject your hypothesis. It's also possible that there may not be enough evidence to decide either way.

There is *no one way* for you to analyze your data. The simplest way would be to look at it and say, "The data seems to support my hypothesis" or, "The data shows that my hypothesis is wrong." This method isn't too convincing. Some way to look at the data mathematically will probably be helpful to you.

One way to analyze the data in the example is to take the *average* heartbeat rate for the taller people and compare it to the average rate for the shorter people.

Here is another way to analyze the data. It's a quick way. Suppose you are still working with heights. Classify each person. Is he in the taller or shorter half of the class? Is he in the faster or slower half of the class? Mark each one *T* (tall) or *S* (short), and *F* (fast) or *Sl* (slow). Now look and see: Are most of the tall people fast (*TF*) or slow (*TSl*)?

There are other ways of treating data which are connected with mathematics. *Graphing* is a very useful method. If you need to review graphing, there is a program on graphing in the Appendix. This program will help you make a graph of your data. The graph, in turn, will probably help you decide to accept or reject your hypothesis.

The graph in Figure 4.9 was made to test the sample hypothesis about height and heartbeat. Does the graph support the hypothesis?

Professional scientists sometimes have difficulty analyzing their data. They are always looking for better ways to do it. You may learn about some of these ways in this course, or you might invent some methods of your own.

One caution! You can never say, "My hypothesis is true," even if the data supports it strongly. You can only say, "My data supports my hypothesis." The reason is that new evidence against the hypothesis might turn up later. Is a hypothesis that seems to be true for your class probably true for people in general?

The data may strongly suggest that the hypothesis is not true. Then you can say with confidence, "My data rejects my hypothesis."

Suppose that your data supports the hypothesis that tall people have fast heartbeat rates. Does this mean that tallness causes fast heartbeat? Or does it mean the reverse? Or is the real reason that the two go together something else, an unknown factor?

The best way to conclude the analysis of your data is to discuss what you found with the whole class and your teacher.

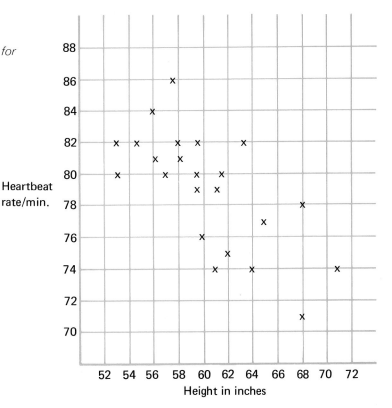

FIGURE 4.9
This graph on height and heartbeat rate provides data for testing hypotheses.

Mastery Item 4-1

How many ways can you classify these students?

a. Look at the students shown in Figure 4.10. State at least four different ways to classify them into two distinct groups based on physical appearance.
b. State four or more ways to separate the people in Figure 4.10 into two separate groups. This time use characteristics *other* than physical appearance.

Organizing information

FIGURE 4.10
What characteristics would you use to classify these students?

Key

a. Many characteristics could be used for grouping these people, for example: height, sex, or hair color. Your teacher should agree that your methods work.
b. You have mastered this part if you stated four classifying characteristics, such as: age, grade level, athletic ability, favorite subjects, number of brothers and sisters, or type of music they like. Again, your teacher should agree with your methods.

FIGURE 4.11
Can you classify these birds into groups which eat the same food?

Mastery Item 4-2

Which birds eat the same foods?

Look at the birds shown on the stamps in Figure 4.11. Which birds might eat the same kind of food?

Key

These groups probably have similar diets: ah, bd, eg, fi. You might also include c with fi. It would probably be incorrect to put a and h together with eg. All four are water birds, but their bills have different shapes.

You may have good reasons for making different groups. You and your teacher should see if your reasons make sense.

Mastery Item 4-3

Why do some animals' hearts beat faster than others?

Look at the data in Figure 4.12. Write two hypotheses that might explain the difference in rates of heartbeat.

Key

There are many possible hypotheses, but each one should have the following characteristics:

a. It is a single sentence only.
b. It could be tested. That is, you can think of a way to get the data needed to support or reject your hypothesis.
c. It should not just summarize the data in different words.

FIGURE 4.12

This data contains normal or average heartbeat rates for different kinds of animals. These measurements come from average adult animals at rest.

Animal	Normal heart rate (beats/min)	Animal	Normal heart rate (beats/min)	Animal	Normal heart rate (beats/min)
cat	180	hamster	347	pig	70
cow	50	horse	45	porpoise	75
dog	100	monkey	220	squirrel	390
goat	85	mouse	610	whale	15

Mastery Item 4-4

Who can run fifty yards the fastest?

Use the data in Figure 4.13 to test each of the hypotheses below. Mark each hypothesis either:

S = supported by the available evidence;
R = rejected by the available evidence; or
I = inconclusive, neither supported nor rejected.

The hypotheses are:

a. Boys can run faster than girls.
b. Older people can run faster than younger people.
c. Brown-eyed people can run faster than blue-eyed people.
d. The heavier a person is, the faster he can run.
e. People can run faster if they are taller.

FIGURE 4.13
Data gathered on 50-yard runners

Subject	Eye color	Sex	Age (yrs)	Height (inches)	Weight (lbs)	Time to run 50 yards (sec)
1	blue	B	12	65	110	6.3
2	blue	G	12	55	100	7.6
3	brown	G	13	64	90	6.2
4	brown	B	13	60	125	7.0
5	blue	B	12	61	120	7.1
6	brown	G	11	65	110	6.9
7	brown	B	12	58	110	7.3
8	brown	G	12	61	90	7.2
9	brown	G	11	56	105	7.5
10	blue	B	12	66	100	6.9
11	blue	G	12	61	90	7.0
12	brown	B	12	60	85	7.1

Key

You have mastered this item if you marked four of the five hypotheses correctly. The best responses are these:

a. **S** or **I**. The average time for the boys is 6.9 seconds and for the girls, 7.1 seconds. Boys seem to be faster, but the difference is small. One more slow boy and fast girl could erase the difference.

b. **S**. The average time for 11 year olds is 7.2 seconds; for 12 year olds, 7.0 seconds; for 13 year olds, 6.6 seconds. A pattern seems to exist.

c. **R**. The average time for each group is just about the same. Because there is no difference, the hypothesis is rejected.

d. **R** or **I**. No method of analyzing the data will show a pattern that supports the hypothesis.

e. **S**. This can be shown several ways. The times for the three tallest can be compared to the times for the three shortest. A graph of height versus time also shows a pattern.

Investigation

5 Describing living things

How can you tell if a living thing is a plant or an animal? In fact, how can you tell if something is really alive?

In this Investigation you will look at different kinds of living things. You will find characteristics they share. You will also look at other things and classify them as living or nonliving, and plants or animals. In these activities you will practice your skills of observing, inferring, classifying, and making hypotheses.

FIGURE 5.1
How could you find out if the things shown on these two pages are alive?

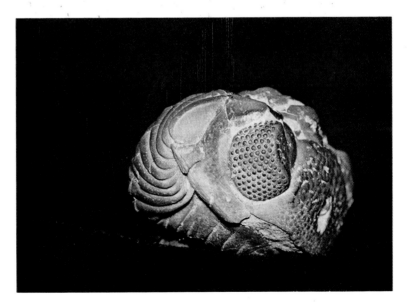

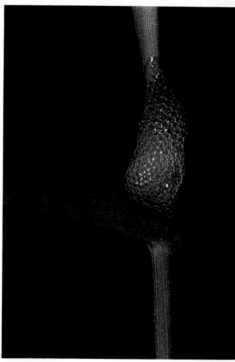

FIGURE 5.2
How could you find out if these things are alive?

Describing living things

Problem 5-1

What does "alive" mean?

Materials
Your teacher will give you things to observe.

Gathering data
Look at the specimens provided for you. Observe them closely. Then, along with your teacher and classmates, try to make up lists of characteristics that separate plants from animals, and living from nonliving things.

Recording data
The lists of characteristics will be written on the chalkboard. Copy in your notebook each list that the class agrees on. If you don't agree on a list with the class, speak out.

FIGURE 5.3
How would you classify these organisms?

FIGURE 5.4
How would you classify the organisms on these two pages?

Mastery Item 5-1

Living or nonliving?

The teacher will give you a new group of specimens. Look at them carefully. Recall what you have just learned about living things. Then try to separate the specimens into two groups: living and nonliving. Write a brief statement telling why you placed each specimen in each group. If you think a specimen cannot be correctly classified, explain why.

Key

Your teacher will discuss your groups with you.

Mastery Item 5-2

Plant or animal?

Look at the organism shown in Figure 5.5. This organism lives in shallow seas. State whether you infer that it is a plant

FIGURE 5.5
Is this organism a plant or an animal? What reasons do you have for your inference?

Describing living things

or an animal, and list the reasons for your choice. What clues can you get from the photograph? If you had one of these organisms in front of you, how would you get evidence to support your inference?

Key

This organism is a sea **anemone** (ah-NEM-ah-nee). This particular anemone lives in the cool sea water off the coast of California. If you called it a plant because it is green, looks like it has petals or leaves, and seems to have a trunk or stem, that would be a good conclusion. If you said it is an animal because it looks like it has a mouth in the middle surrounded by arms, this would also be a good conclusion. If you said you didn't have enough data to tell whether it was a plant or an animal, that was probably the best response. The anemone is really an animal, but it is green because it has tiny plants inside it.

FIGURE 6.1
Do all living things breathe?
What about this hamster?

Investigation

6 Designing experiments

In Describing Living Things, did your class suggest that all living things breathe? When you breathe, you take air into your body and then let it out. You know that. Actually, the air you take in and let out is not the same.

Part of air is a gas called **oxygen** (OX-uh-gin). This is the only part of the air that you use. The oxygen you breathe helps turn food into energy. The food and oxygen both change:

food + oxygen $\xrightarrow{\text{turns into}}$ energy + carbon dioxide + other products

Your body does not use the new gas, **carbon dioxide** (CAR-bun dy-OX-ide) and you breathe it out. By breathing, you exchange gases with your **environment** (en-VY-run-ment), or your surroundings. The environment is changed a little each time oxygen and carbon dioxide are exchanged.

Do all living things exchange gases with their environment? You can observe people breathing, but what about other living things? Do plants also exchange gases with the environment? Do animals that live in the water exchange gases? How can you tell whether an organism is giving off or using up a gas? In this Investigation you will experiment to answer some of these questions.

Your experiences in this Investigation will also help you to tell a good experiment from a poor one.

Problem 6-1

Do aquatic plants and animals exchange gases?

Sometimes it is difficult to tell that an organism is using up oxygen. It is easy to tell if an **aquatic** (ah-KWAH-tik) organism is giving off carbon dioxide. Aquatic organisms live in the water. In your experiment, you will try to detect carbon dioxide. If you detect it, you can assume that the organism is using up oxygen. The amount of carbon dioxide given off should be the same as the amount of oxygen used.

FIGURE 6.2
The liquid in the bag turns from blue to yellow when carbon dioxide is added. What happens when you breathe into the blue water?

Materials

test tubes
plastic bag or other container
soda straws
blue water
snails
water plants

Gathering data

The blue water you will use in your experiment contains a chemical. If you add carbon dioxide to it, the chemical turns yellow. You can test the blue water before you do the experiment. What should happen when you breathe out through the straw into the blue water? Test your prediction. Use the plastic bag, the soda straw, and the blue water as shown in Figure 6.2. Did everyone get similar results?

Hold the plastic bag firmly around the top and shake it. Make the liquid mix with the air in the bag. Watch for color changes. Can you shake the carbon dioxide out of the water?

Now you can investigate the question, "Do snails and water plants take in oxygen and give off carbon dioxide?"

Look at Figure 6.3. Each group will prepare a set of tubes like this. Try to choose the same size plants and animals for each tube. Also, put the same amount of blue water in each tube. Then make sure the tops or stoppers are tightly sealed. Mark your tubes so you can identify them.

FIGURE 6.3

These test tubes are filled with the blue water. Snails are placed in two of the tubes, and water plants in two others. Then half the tubes are placed in the dark. What do you think will happen in each tube?

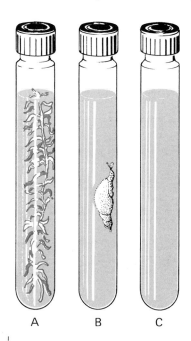

A B C D E F

Place one set of tubes in a light part of the room, but not in direct sunlight. Place the other set in a dark place such as a cupboard or closet. Tomorrow you will compare the two sets of tubes.

Each tube should answer a question about gas exchange. Make a table in your notebook like Figure 6.4 and write in the question you think each tube will answer.

Recording data

Look carefully at the color of the water in each tube. Record any changes that occurred. Then reverse the position of the two sets of tubes. Set 1 now goes in the dark and Set 2 in the light. Record the color of the water again after 24 hours.

Analyzing data

What does your data tell you about the gases given off and used up by snails and water plants? Keep in mind the data you have from blowing into the blue water and then shaking

FIGURE 6.4

Use a table like this one to record your data for gas exchange tests.

	Tube	Question	Color	Answer
First day	A	Do plants give off carbon dioxide in light?		
	B			
	C			
	D			
	E			
	F			
Second day	A			
	B			
	C			
	D			
	E			
	F			

it. How does the data answer the questions you wrote in your table? Look especially at tubes C and F. What did they tell you?

An experiment usually has a **control.** A control is for comparison. It should make you feel sure that you are testing only one factor in your experiment. A control is set up exactly like the experiment, except it leaves out the factor you are testing. What controls, if any, did you have in your experiment on gas exchange?

Problem 6-2

Measuring gas exchange

Look at the **manometer** (ma-NOM-eh-ter) in Figure 6.5. With this apparatus your group will measure how fast an organism uses oxygen. You learned in Problem 6-1 that when an organism uses up oxygen, it gives off carbon dioxide. Notice the bag of chemical pellets in the manometer. The pellets absorb carbon dioxide. So as the organism uses up oxygen, the amount of gas in the container is reduced.

Gases in the manometer are used up and absorbed. Then the air pressure inside the bottle is less than outside. The

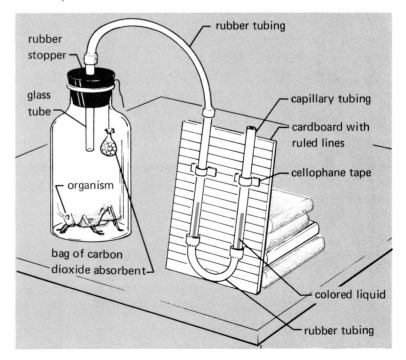

FIGURE 6.5

This apparatus is a manometer. It can measure how fast an organism uses up oxygen.

difference in pressure pushes the colored liquid higher up in one side of the tubing. Look at Figure 6.6a. At the beginning of an experiment the air pressure is the same on both sides of the liquid. Later in the experiment (Figure 6.6b) there is less air pressure on the liquid in the tubing on the left. The liquid is pushed up by outside air pressure. Drinking through a soda straw works about the same way.

Materials

animals	manometer
plants	balance

Gathering data

Put the parts of the manometer together as shown in Figure 6.5. The teacher will also give you directions. **CAUTION:** The carbon dioxide absorbent is a dangerous chemical. If you touch it or get some on your clothes or on the experimental animal, be sure to tell the teacher.

FIGURE 6.6
Watch the height of the liquid in the tubes of the manometer. The level changes when air pressure changes.

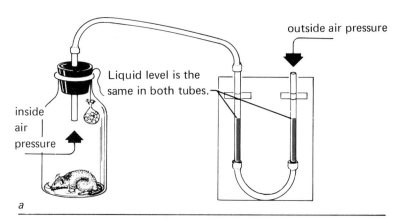

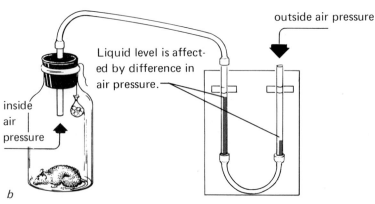

Designing experiments

Now your group should decide what your experiment will be. Your teacher has provided organisms you can use in this Problem. Or perhaps you have a small pet you might like to use, instead of the organisms in the room. If so, ask your teacher. You can use the manometer and live organisms to ask questions like these.

a. How far will a frog make the liquid move in five minutes?
b. Does a snake use as much oxygen in 10 minutes as a frog does?
c. Do seeds use oxygen?

Choose one of these questions to investigate or think of one yourself. Write down your questions and check with your teacher before you begin.

Remember the role of Tubes C and F in Problem 6-2? These tubes showed that the blue solution stayed blue when no organisms were added. Those tubes were your experimental controls. What factor will you be testing in this experiment? How can you be sure that a change in the

FIGURE 6.7
How quickly does a snake use up oxygen?

FIGURE 6.8
Does a grasshopper use oxygen?

FIGURE 6.9
Does a 125 g frog use 5 times as much oxygen as a 25 g frog?

level of colored liquid is due to the test organism—not some other factor? Your group should decide what control to have.

To use the manometer, place the experimental organism in the bottle and put the stopper in firmly. The level of the liquid in both tubes should be the same when you begin.

If the animal you use can bite, ask the teacher the best way to put it into the bottle. Usually you can get the animal out if you put the bottle on its side and let the animal come out by itself. If for any reason the absorbent in the bag gets wet, call your teacher. Drippings from the absorbent will hurt your test organism.

The colored liquid in the tube should move soon after you begin your experiment. If it does not, then your system may leak. To check for a leak, push down gently on the stopper. The level of the liquid should change. If it does not, then the system probably leaks. Just pushing firmly on the stopper often will stop a leak. If you suspect there is a leak, put petroleum jelly around the stopper.

Make several observations for each experiment. You can record the movement every 30 seconds, one minute, or five minutes. It depends on how rapidly the liquid is moving. If you have time, it is a good idea to repeat an experiment to be sure of your results. When you do this, loosen the stopper between experiments. This will give the organism a fresh supply of oxygen. It will also return the level of the liquid to the starting point.

Recording data

Keep accurate records of your data. A table might be a good way to record your data. Write a brief report that shows how your data answers your question.

Analyzing data

Discuss your data and the control you used with other experimenters in the classroom. Can you make any general statements about gas exchange in living things? Did active animals use more oxygen than ones that sat still? Do larger animals use more oxygen than small ones? Do seeds use oxygen? Can you think of a way to find the amount of oxygen used by one seed? Can you think of a better control than the one you used?

Mastery Item 6-1

A survival experiment

Look at Figure 6.10a. This shows you the beginning of an experiment that was set up by a junior high school student. This student was trying to find out if plants and animals could live in closed containers.

Figure 6.10b shows the same set up one week later. What do you think is the best explanation for the results of this experiment?

Key

The plants were necessary to provide enough oxygen to keep the animals alive. Without plants, the animals died. With plants, they survived.

FIGURE 6.10

A survival experiment. Figure 6.10a shows the beginning of the experiment. One week later the setup looked like Figure 6.10b. What do you think happened?

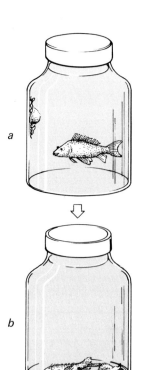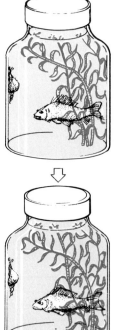

Mastery Item 6-2

Do roots need oxygen?

This Mastery Item is designed to see if you can tell a poor experiment from a better one. Look at the two experiments pictured in Figure 6.11. Two students hypothesized that the roots of plants need oxygen to grow. To test their hypotheses, each student grew tomato plants in jars of water. They put in chemicals for healthy plant growth.

The first student carefully bubbled oxygen for two weeks through the water all his plants were growing in. He used a tank of oxygen to do the job. The second student used a similar oxygen tank. He bubbled oxygen through a jar with half his plants in it. The other half he did not treat. Each was careful to give the plants equal amounts of light. All the plants in both experiments lived and grew. Both students gathered all of the data they wanted. Whose experiment would you consider a better test of the hypothesis?

Key

The second student designed a better experiment because he could compare his treated plants with untreated ones. Then he would know if the oxygen made a difference. The first student had no control with which to compare his treated plants.

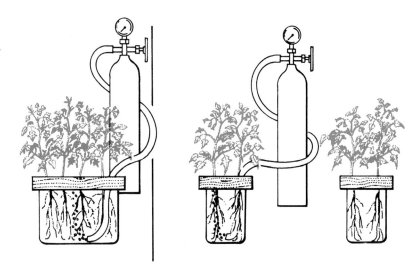

FIGURE 6.11

These two setups were designed to find out if roots need oxygen.

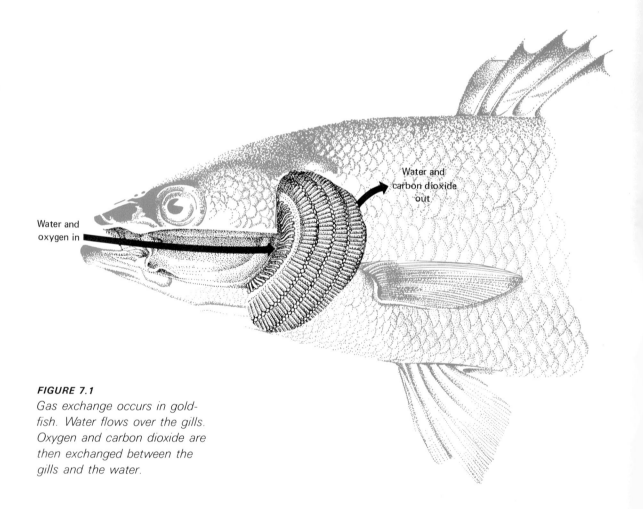

FIGURE 7.1
Gas exchange occurs in goldfish. Water flows over the gills. Oxygen and carbon dioxide are then exchanged between the gills and the water.

Investigation

7 Investigating behavior

In this Investigation you will be concerned with changes in the behavior of goldfish. You will have a chance to make hypotheses about factors that might cause goldfish behavior to change. Then you will design experiments to test your hypotheses. In this way, you will understand more about how scientists work.

In Investigations 5 and 6 you discovered ways in which living things interact with their environment, or surroundings. For example, plants and animals exchange gases with their environment. You know that water animals also exchange gases with their environment. To breathe in a watery environment usually takes different bodily organs.

Study Figures 7.1 and 7.2 to become familiar with the behavior of the goldfish as it breathes.

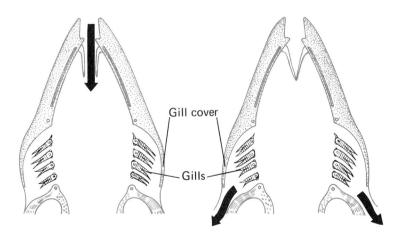

FIGURE 7.2
How a goldfish breathes. Figure 7.2a shows how the fish takes in gulps of water containing gases. Its mouth is open, and the gill covers are closed. Figure 7.2b shows the goldfish pushing out water through the open gill covers. Its mouth is now closed.

After studying these Figures, ask your teacher for a goldfish. Observe the fish and study its behavior as it breathes. How is the goldfish's behavior like your breathing?

Problem 7-1

The goldfish puzzle

The following story describes a puzzling change in goldfish behavior.

One winter morning, a boy living in northern United States was given a goldfish for his birthday. He placed the goldfish in a bowl on the sill of a bright, sunny window.

That afternoon before the sun had gone down, he sat down to watch his fish swimming. He noticed that the fish seemed to be active and was breathing rapidly.

Early the next morning when the boy woke up, it was just becoming light. Heavy snow was falling. The boy looked at his fish again. It still looked healthy, but its behavior puzzled him. He thought his fish now seemed to be breathing very slowly.

What do you think caused the behavior of the fish to change? Read the story again and look for clues to the problem. Suppose you could ask the boy three questions to get more information about the problem. What would you ask?

Materials

Ask your teacher for the materials you think you will need for your experiment.

Gathering data

You now have a chance to conduct experiments to help solve the goldfish puzzle.

Write down some hypotheses about what caused the fish's behavior to change. Work with several of your classmates and design an experiment to test at least one of your hypotheses. Write a brief description of the experiment you propose to do.

Check with your teacher before you carry out your experiment.

FIGURE 7.3

How would you investigate the change in the fish's breathing rate?

Recording data

Record any observations and data from your experiment in a table that you design.

Analyzing data

Organize your data in any way that will help you interpret it. Does your data support your hypotheses? What conclusions can you draw from your data? Write down your conclusions. Do you now have a possible solution to the goldfish puzzle?

Suppose you decide to change the experimental conditions even more than you did. Examine your data again. Would it help predict how fast the fish would breathe under the new condition?

A graph often helps make sense of data. It is a kind of diagram of data. Graphs can also be used to make predictions. If you do not know how to graph data, turn to the Appendix. Do the program there, How to Graph Data, and you can teach yourself how to make a graph. Try graphing your goldfish data. An example of a graph is shown in Figure 7.4. If you connect the points, you will have a line graph.

In class discussions and other activities, you will learn more about how useful line graphs are.

FIGURE 7.4
Turn to Part B of the Appendix to find out how to make a graph like this.

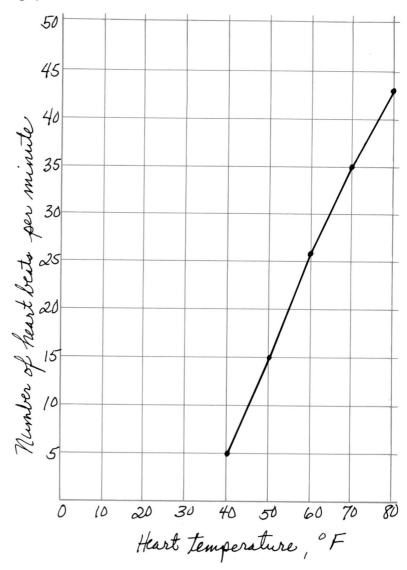

Mastery Item 7-1

What makes flies thirsty?

A biologist, who spends much of his time studying the behavior of a common fly, once asked, "What makes a fly thirsty?" He knew what made people thirsty. He hypothesized that the same factors would make flies thirsty. He designed several experiments to test his hypotheses.

Your task is this.

a. Read and analyze the descriptions of his experiments.
b. Write down the hypothesis you think was being tested in each experiment.
c. Tell if the data supports each hypothesis and give reasons for your answers.

EXPERIMENT 1

The biologist set up six groups of 100 flies each. All the flies were the same kind. He gave one group of flies plain water. The other groups he fed water mixed with different amounts of salt. The experiment lasted 24 hours. Figure 7.5 shows how he organized the data he got.

EXPERIMENT 2

The biologist kept groups of 100 flies in drying chambers for different periods of time. A drying chamber is a special

FIGURE 7.5
Data on salt experiment

Weight of salt in 100 ml water (grams)	Milliliters of water that 100 flies drank in 24 hours
0	1.21
5	1.20
10	1.18
15	1.21
20	1.20
25	1.21

FIGURE 7.6

Data on dry air experiment

Number of hours in the drying chamber	Milliliters of water that 100 flies drink in 25 hours
0	1.20
5	1.26
10	1.32
15	1.41
20	1.47
25	1.53

container. A chemical in it readily absorbs water from the air. The air then becomes very dry. Another 100 flies were kept in a container where the air was not dried. All the flies were the same kind. The data he got is shown in Figure 7.6.

Key

Experiment 1 is designed to test this hypothesis: Salt makes flies thirsty. The flies don't seem affected. A change in the amount of salt does *not* change the amount of water they drink. The flies given salt drink about as much as the control group. Therefore, the hypothesis is disproved.

Experiment 2 is designed to test this hypothesis: Dry air makes flies thirsty. The data shows that flies in dry air *do* drink more than the control flies. Therefore, the data supports the hypothesis. The flies in the drying chambers probably sweated a lot. That would have made them thirstier than the control flies.

Mastery Item 7-2

What factors affect your pulse rate?

You know that your pulse or heartbeat rate changes from time to time. But do you know what factors or conditions cause these changes?

Your task is this.

a. Make a hypothesis to account for changes in your pulse rate.

FIGURE 7.7
Does the exercise this student is performing affect her pulse rate? How much?

b. Design and conduct an experiment to test your hypothesis.
c. Interpret your data. Does it support your hypothesis?
d. Make a record of your entire investigation.

Key

Your investigation was successful if:

a. Your hypothesis could be tested.
b. Your experimental design included a control.
c. You could tell from your data if your hypothesis was supported.
d. Your description of the experiment was clear and complete. (A classmate should be able to read your description and perform the same experiment.)

A good experiment does not depend on picking a factor that does cause changes in your pulse rate. The important thing is to be able to *test* a hypothesis. It can be just as interesting to learn what does *not* affect pulse rate.

Mastery Item 7-3

How does exercise affect your pulse rate?

In this Item your job is to make predictions from graphed data. You need a watch or clock with a second hand. Stand still for two minutes, then have a classmate count your pulse rate for 15 seconds. Record your data in a chart like Figure 7.8.

After the first count is recorded, do three deep knee bends without stopping. Your partner must count your pulse *immediately* afterwards. Otherwise you will miss the effect of the deep knee bends.

FIGURE 7.8
Record your data on pulse rates and deep knee bends on a table like this one.

Date: _____
Name of subject: _____
Name of assistant: _____

Number of deep knee bends	0	3	6	9	12	2	8	14			6
Pulse rate per 15 seconds									22	28	

Investigating behavior

Stand in one spot and rest. Wait until your pulse rate returns to about the first count you made. Then do six deep knee bends without stopping and have your pulse counted immediately. Rest again.

Repeat the procedure with nine, then twelve deep knee bends. Now graph your data. Use your graph to help make the predictions asked for below. Use dots to show your data and use "X's" to show predictions.

a. Predict your pulse rate after two, eight, and fourteen deep knee bends. Put an "X" on the graph for each predicted rate.
b. Predict how many deep knee bends will make your pulse beat 22 times in 15 seconds; 28 times in 15 seconds. Put an "X" on the graph for each prediction.
c. Predict your pulse rate after doing six deep knee bends again. Put an "X" on the graph for your prediction.
d. Write down why you put the "X's" where you did.

Now collect data to test your predictions. Record the data in the data table and as dots on the graph. How accurate were your predictions? Tell why you think that your predictions are either confirmed or not confirmed.

Key

Your investigation was successful if:

a. You gathered data that shows a faster pulse rate after exercise than after rest.
b. You properly constructed a line graph of the data (check with your teacher).
c. You based your predictions on the pattern of the graph.
d. You gathered the necessary data to test your predictions and marked it on the same graph.
e. You said the predictions are confirmed when the new data fits the pattern of the graph. Predictions are not confirmed when data does not fit.

FIGURE 8.1
Which of these living things can learn?

Investigation

8 Learning

What is learning? Have you learned anything today? Can animals other than humans learn? Can simpler animals like earthworms learn anything? Can all living things learn? How can you tell whether an animal has learned something?

In this Investigation you will answer some of these questions and others that you will ask yourself. Even if you think you can answer some of them now, you will know more about learning after doing this Investigation. You will learn how learning can be studied scientifically.

FIGURE 8.2
What kinds of things does a baby have to learn?

FIGURE 8.3
This boy has learned to toss newspapers onto front porches. Or has he? How could you measure how well he has learned this skill? How could you tell whether he has learned the skill better a month from now?

Learning

Learning is much more than what you do in school. You may have learned more before you were five years old than you will learn in the rest of your life. Many things you now do without thinking, you once had to learn through trial and error. What are some of the things you have learned?

Learning can be studied scientifically. However, a big problem is to decide what kind of learning to study. Things like learning to play baseball are difficult to study because they are complicated. How many kinds of things do you learn in baseball? It will be easier to study simple types of learning where controls can be used.

In this Investigation you will begin by studying maze learning because it is quite simple and interesting. Many different animals, including humans, can learn mazes.

Problem 8-1

The finger maze

Have you ever watched a classmate studying? Was he learning something at the time? How could you tell?

The only way of knowing that a person has learned something is for him to show it. For instance, your singing a song proves that you have learned the words. Just telling someone that you had learned to ride a bicycle might not convince him. But riding past him would prove that you had learned.

FIGURE 8.4
What has she learned?

FIGURE 8.5

How could you tell if this rat has learned to go through the maze?

Suppose a friend made one out of 20 basketball shots. Would you say that he had learned how to play basketball? Your answer is probably both yes and no. You might say that he had learned a little, but not much. There was still room for improvement or more learning.

Imagine that a month later you saw your friend make 10 out of 20 shots. You might infer that during the month he learned something. Or maybe he was just lucky this time. But if he could make 10 out of 20 shots a second time, you would probably think that it was more than luck.

In this Problem you will see how learning can be inferred from a person's performance in a T-maze. Figure 8.5 shows a rat in a Y-maze. You can see how the maze gets its name. The passageways look like a series of Y's.

When a rat goes through a maze, it is fed when it gets to the end. The food is a reward or prize. It gives the rat a goal at the end of the maze. Sometimes people will learn a maze just because learning is fun and challenging.

Materials

maze for human subjects (You may help make it.)
stopwatch or clock with a second hand   blindfold

Gathering data

Choose a partner. One of you will be the *experimenter* and the other will be the *subject*. If there is enough time, you

Learning

can change jobs after the first set of data is obtained, and repeat the experiment.

Help blindfold the subject and have him sit at a table or desk. He should not see the maze during the experiment. The experimenter puts the maze in front of the subject and holds it so it doesn't slide around.

If you are the experimenter, place the subject's pencil or finger at the starting point, as shown in Figure 8.6.

When you are ready to start timing, tell the subject to "START" through the maze. When the subject reaches the end of the maze, say "STOP." Write down the time it took.

Return the subject's pencil or finger to the starting point. Do a second trial. Don't tell the subject how well he did or how much time he took. The subject should do at least six, and preferably 10, trials. If the subject and experimenter later trade jobs, use a different maze. Why?

Recording data

The experimenter should record the number of each trial and the time the subject takes to go through the maze. The record should note also if anything unusual happens during a trial. An example of such a note might be, "Subject dropped pencil," or "Subject got out of maze and had to be put back in."

Analyzing data

To help you analyze your results, it will be a good idea to plot your experimental data in a graph. The numbers of the trials should be plotted on the horizontal axis. The time required for each trial should be plotted on the vertical axis.

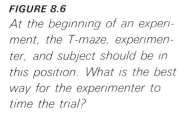

FIGURE 8.6

At the beginning of an experiment, the T-maze, experimenter, and subject should be in this position. What is the best way for the experimenter to time the trial?

Before the class discusses the results, you may want to look at the graphs that others made. Are the other graphs similar in shape to yours?

The questions below may help you to analyze your experiment and the graphs you have seen.

a. What makes you think that you learned during the experiment?
b. How did you learn to do the maze? What skills or clues did you use?
c. If you did 20 trials, what would you predict your time would be on the twentieth?
d. What would the graph look like if you had learned the maze more slowly than you did? More rapidly?
e. What do you think a learning graph might look like if you repeated the experiment with the same maze tomorrow? What if a different maze were used?

Problem 8-2

The mouse in the maze,™ a simulated experiment

In Problem 8-1 you got your data from a real experiment. In this Problem you will get data without using live subjects directly. This is called a **simulation** (sim-yoo-LAY-shun).

A simulation is similar to the real thing. You may know about some simulations. One is a flight simulator in which airplane pilots first learn how to fly. A flight simulator looks like the cockpit of an airplane. It has a full set of controls, which are connected to a computer. The student pilot operates the controls. The computer tells the pilot what a real airplane would probably do in the situation. The flight simulator doesn't really fly, but it responds to the student's operation of the controls in a realistic way.

Some high school driver training classes use automobile-driving simulators. You also have probably heard of computers playing chess or tick-tack-toe. The computer simulates a person's thinking. It plays these games as if it were a real person.

Instead of using a computer in this Problem, you will use a set of special cards called "simulator cards." They can be used like a computer to get realistic experimental data.

Imagine that you have equipment to construct a T-maze similar to the Y-maze in Figure 8.5. You can make the maze with any number of T's in it. Imagine also that you have lots of small animals (rats, mice, and hamsters) to put in the maze. You also have an imaginary laboratory assistant who will quickly do any experiment you want.

The assistant can measure the time it takes an animal to go through the maze. He can also count the number of wrong turns (mistakes) that the animal makes in each trial. You can probably think of several learning investigations to do with such an outfit. Describe briefly one or more investigations you might like to try. What hypothesis will you be testing?

The simulator cards can probably be used to investigate these questions. The cards will give data like the data from real experiments. You will still have to analyze the data yourself.

FIGURE 8.7
It's lucky this pilot is only experimenting with a simulation flight.

FIGURE 8.8
What questions could you investigate by putting these animals in a maze?

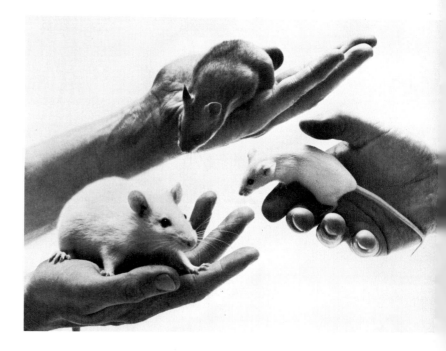

Materials

set of simulator cards, "The Mouse in the Maze"™

Gathering data

Choose the partners you will work with as a team.

Get a set of simulator cards from your teacher and look it over. The colored cards with holes punched in them stand for experimental conditions. You pick different cards to choose different conditions. You can choose the number of T's in the maze, the kind of animal in it, and the number of the trial in the experiment.

There is also a set of data cards on which numbers are printed. One side is marked "Time." It will give the time, in seconds, taken for one trip through the maze. The other side is marked "Mistakes." It will give the number of mistakes, or wrong turns, made in one trip. You can measure the animal's performance in either time or mistakes.

In the game you can change some of the experimental conditions, but other conditions you can't change. For instance, the temperature is 25°C, and the maze is lit as brightly as a classroom. Each animal receives a small amount of food at the end of the maze. None of the animals has been in a maze before, and trials are made one hour apart.

Learning

HOW TO USE THE SIMULATOR CARDS

a. Select one green card. It tells which animal is being studied.
b. Select one orange card. It tells the number of T's in the maze.
c. Select one blue card. It tells the number of the trial for the animal in the maze.
d. Fit the three colored cards together. One hole should go through all three.
e. Select one side of the data cards to use, either "Time" or "Mistakes."
f. Roll the die. The number on the die tells the number of the data card you should use. The number will be from one to six.
g. Place the three colored cards over the data card you just picked. Line up all the cards.
h. Read the one number that shows through the hole in the colored cards. This number is either the number of seconds the animal took to run the maze or the number of errors it made, depending on whether you chose "Time" or "Mistakes."
i. Repeat steps a—h as often as necessary. Be sure to roll the die to determine a new data card for each trial.

Do as many trials as you think you need to provide data for a learning graph. If you have enough time, you may want to conduct another experiment. You might investigate different animals or different mazes.

Recording data

For every simulated experiment that you do, record the conditions and outcomes. An example of a useful record of data is shown in Figure 8.9. Your data record doesn't have to be exactly the same, but it should keep your data organized.

Analyzing data

Make a learning graph using the data from your experiment. If you did more than one experiment, make a graph for each experiment.

Use your learning graph to answer the questions that follow. Be ready to discuss your answers with the class. Remember that you can talk about the data from simulated

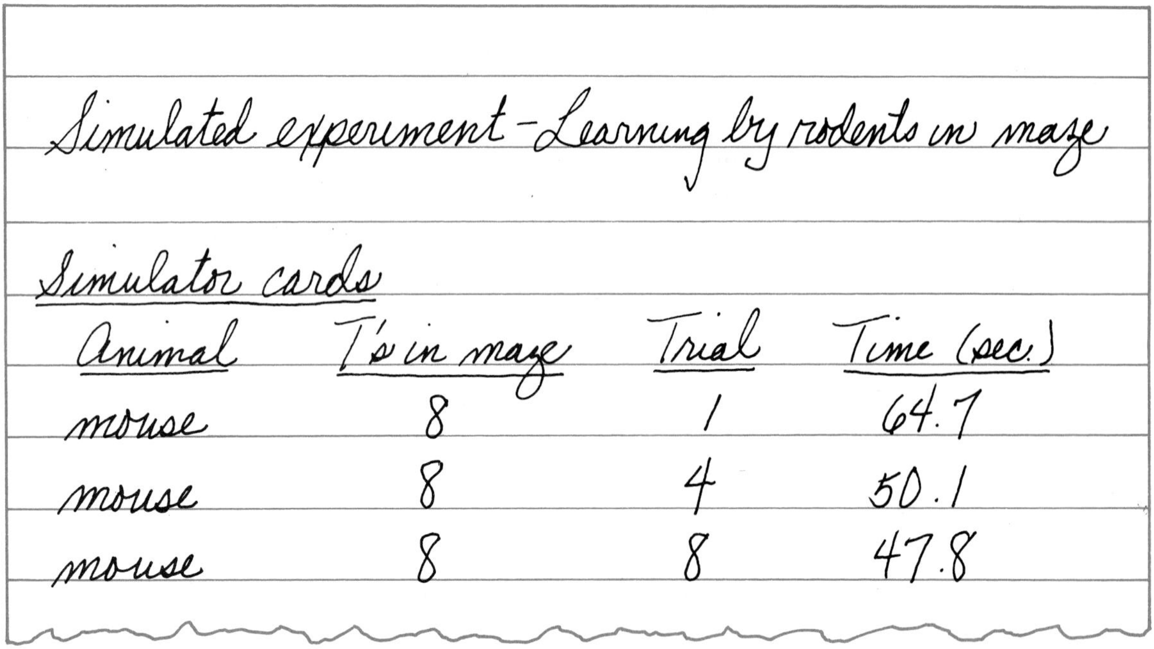

> *Simulated experiment – Learning by rodents in maze*
>
> Simulator cards
>
> | Animal | T's in maze | Trial | Time (sec.) |
> |--------|-------------|-------|-------------|
> | mouse | 8 | 1 | 64.7 |
> | mouse | 8 | 4 | 50.1 |
> | mouse | 8 | 8 | 47.8 |

FIGURE 8.9
A student record of a simulated experiment in maze learning.

experiments as if it were real. You may want to think back to Problem 8-1 to answer some of these questions.

a. Did the animals learn? How can you tell?
b. Did all the animals learn at the same rate?
c. What similarities and differences can be found in maze learning by humans and other animals?

Problem 8-3

Can animals learn to escape?

Your investigation of learning has been limited to maze learning. There are many other kinds of learning, such as learning to escape or get out of something. Maybe as a young child you learned how to get out of your crib, playpen, or swing.

In this Problem you and your class will perform an experiment to find out if small animals can learn to escape from a flower pot.

Materials

small, hungry animal such as a mouse, hamster, rat, or gerbil
flower pot with unglazed walls

Gathering data

In this Problem the whole class will probably work together. Place the flower pot in the middle of a large table or desk, and surround the pot with small bits of food. Put the animal inside of the flower pot, and measure the time it takes to escape.

Let the animal eat only *one* bit of food. Put the animal back in the pot, and again measure the time for it to escape. Be gentle with it.

During these experiments try not to bother the animal by getting too close or making loud noises. Why? Repeat the experiment at least six times.

FIGURE 8.10

How can the experimenter keep the gerbil from escaping from the lab table?

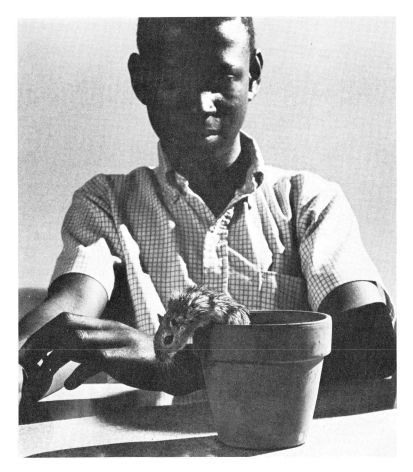

Recording data

Keep a record of the time required for the animal to escape each time. Make notes of any interesting behavior by the animal during the experiment.

Analyzing data

Make a learning graph from the data obtained by your class.
  Discuss the results of your experiments by starting with these questions.

a. Did the animal show signs of learning?
b. How does this learning graph compare to the ones obtained in Problem 8-2?
c. What other animals could be tested for learning in an escape situation? What equipment and procedures would you use?

FIGURE 8.11
Does listening to the radio really interfere with learning?

Mastery Item 8-1

Studying with the radio on

"Turn off the radio! How do you expect to learn anything while it's on? Get your studying done first, *then* listen to the radio if you want to!"

These words have been heard by thousands of teen-agers. The speaker has the idea that learning is prevented by listening to a radio. Describe an investigation to test the idea that listening to the radio interferes with learning.

Key

Your investigation should contain a plan for controlling many factors. One way to do this is to compare a subject who listened to the radio while studying, with another person who did not listen to the radio. However, how could you be sure that any difference was not caused by another factor, for example:

a. One person was better at the subject studied than the other.
b. One person tried to learn something more difficult than the other.
c. One person studied in a warmer room.

The more variable factors you control in your plan, the better it is. Your investigation could use the same subject in two trials. In one trial he would study with a radio on. In a control trial the radio would be off. How could you be sure that differences weren't due to the study material?

Mastery Item 8-2

Learning without reward

Two groups of rats were tested in a Y-maze once each day. A Y-maze is built like a T-maze. This one had 14 Y's.

There were eight rats in each test group. Rats in group A were rewarded with a small amount of food when they got through the maze. Rats in group B were *not* rewarded. Figure 8.12 shows the results of a series of trials on groups A and B. Notice that the time plotted for each trial is the average for the group of eight, not an individual score.

Answer the following questions and give reasons for your answers.

FIGURE 8.12

Maze performances by rewarded and unrewarded rats.

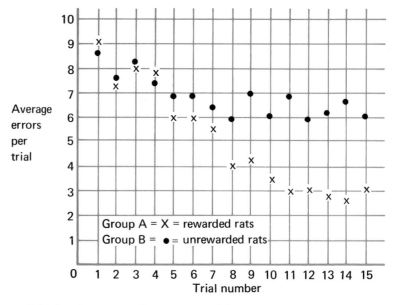

a. Did both groups learn something?
b. Which group learned more?
c. Which group learned faster?
d. What results would you expect from Group A if the same experiment were continued for 50 trials?
e. What hypothesis can you make to explain why both groups seem to do equally well at first?
f. Does this experiment sound like anything that has ever happened to you?

Key

a. Yes. The downward slope of each graph is evidence of learning.
b. The rewarded group probably learned more since the difference between its starting performance and final performance is greater.
c. The steeper slope of the rewarded group indicates that it learned faster than the unrewarded group.
d. The graph would probably flatten soon, and there would be no further change with more trials. However, there is a possibility that a slight improvement would occur.
e. Many reasonable hypotheses are possible. One might have to do with the curiosity of the animals. That is, at first the rats are driven mainly by curiosity to reach the end of the maze. After the first few trials the unrewarded rats do not have the added interest of food, which the rewarded rats do.
f. Did you think of any cases where you were rewarded for learning something? Any cases where you learned just because you wanted to? Any cases where you stopped learning after awhile? Why?

Mastery Item 8-3

Can ants learn?

Invent and describe a task, other than going through a maze, that might be used to study whether ants can learn.

Key

You have mastered this item if the task you described:
a. Involves some action of the ant that can be observed.
b. Includes a measurement of the action of the ant.
c. Can be done by ants; that is, it is practical.

Unit **two**

The environment affects living things

What would you do if a flood or an earthquake destroyed your home? What could you do? The homes of many living things change constantly. Rainwater floods earthworms out of their tunnels. Windstorms blow down trees. Fiddler crabs live in the edge of rising and falling tides. The picture on the left is a flooded cornfield.

Each living thing is affected by everything around it. The study of how the environment affects living things—and how they affect their environment—is called **ecology** (ee-KOL-uh-gee). Some of the ways an environment affects plants and animals are easily seen. One example is wind blowing down a tree. Many environmental effects are more difficult to see. It is possible to set up experiments to investigate these effects. That is what you will do in Unit Two.

You will learn to do these things and some others:

a. Design experiments to find out how the environment affects plant and animal behavior.
b. Predict some of the characteristics that plants and animals need to survive in new environments.
c. Discuss the factors that seem necessary to maintain life on earth.

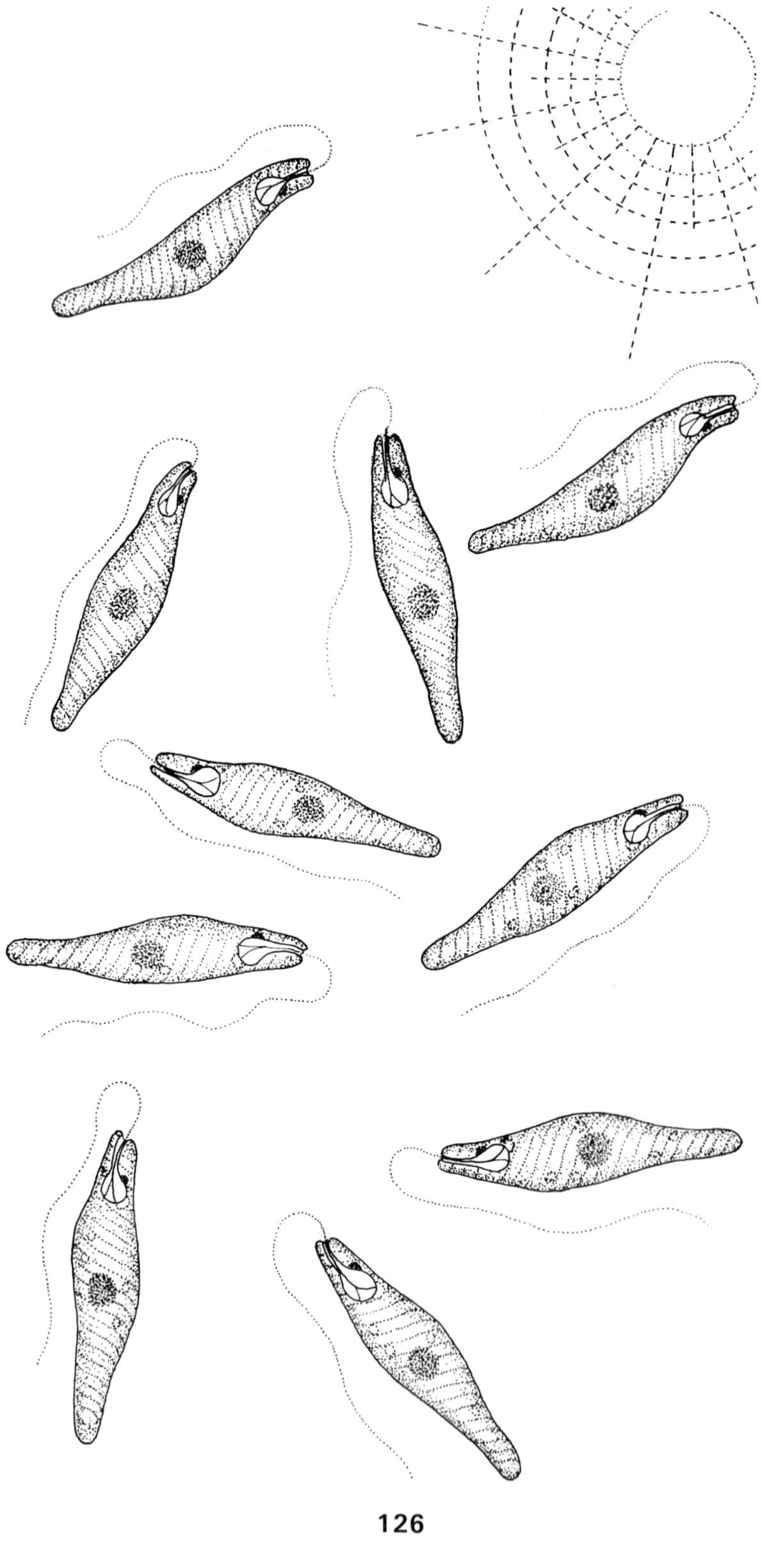

FIGURE 9.1
How do euglenas react to light?

126

Investigation

How does light affect euglenas?

An organism must adjust to changes in its world, or it will die. In this Investigation you will learn how **euglenas** (you-GLEE-nuhs) respond to changes in light. You will also infer why their responses help them to survive. (See Figure 9.1.)

Euglena is the name of a group of microscopic organisms. The individuals are simply called euglenas. If you have observed water from a pond or ditch through a microscope, you may already have seen them.

Your teacher has been growing a population of euglenas for you to study. They are living in a thin "pea soup" made from water and ordinary split peas.

The soup is green because it contains **chlorophyll** (KLAWR-uh-fil). Chlorophyll is the green substance that gives all green plants their color. Chlorophyll can absorb light energy from the sun. Green plants then use this energy to make their own food.

On a calm summer day the green plants in a pond look as if they are resting. But is this really so? What did you discover in Unit I about green plants?

Water and minerals are being drawn silently but steadily from the pond and soil at the bottom. Leaves are exchanging gases with the water or air. At the same time the chlorophyll in the leaves is absorbing the energy from sunlight. The plants use this energy from sunlight to turn water and carbon dioxide into high-energy foods.

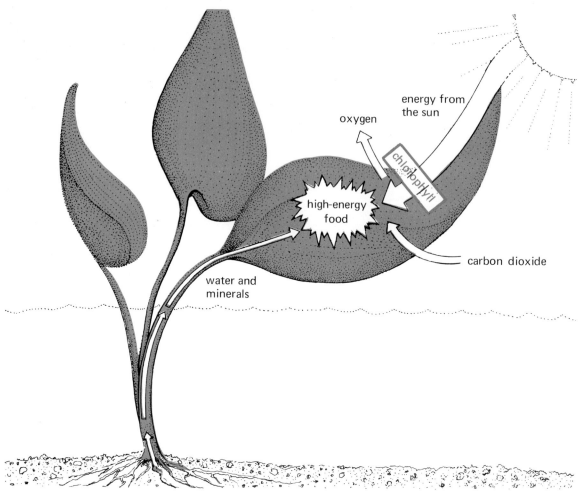

FIGURE 9.2
A plant can transform the sun's energy into food energy. Water and minerals are brought up through the roots, and water escapes through the leaves.

The manufacture of food by plants is a most important activity. It makes life on earth possible. Almost all other organisms eat green plants or eat animals that eat green plants. The animals you eat once depended on plants for their food and energy. The manufacture of food by plants is important for another reason. Oxygen is given off during the process. In fact most of the oxygen that organisms use comes from plants with chlorophyll. (See Figure 9.2.)

Look again at the "pea soup" the euglenas are in. What is the source of chlorophyll in the soup? Did it come from the peas? Is it in the euglenas? You will need a microscope to look at the soup and find out. There are directions for using a microscope in the Appendix.

First prepare a wet mount from the soup. Examine your slide under the lowest magnification of your microscope.

You should see many euglenas swimming about. To observe a euglena more closely, use the highest magnification of your microscope. Refer to the Appendix to see how you change the magnification.

Try to discover how a euglena moves and which end usually goes first. Figure 9.1 may help you to see more of the euglena's parts.

Can you see a reddish-orange spot toward one end of a euglena? If not, try reducing the amount of light slightly with the diaphragm of the microscope. A long whip or **flagellum** (fluh-JEL-um) is attached to the euglena next to this spot. There is a lump or swelling on the base of the flagellum. This lump contains substances which seem to be sensitive to light.

Continue making observations. Could you accurately describe a euglena to someone who has never seen one? That's a good test of how careful your observations are.

Can you now answer these questions:

a. Where is the chlorophyll located in the soup?
b. How do you think euglenas get their food?
c. Most organisms that contain chlorophyll cannot move about. What advantage could it be for euglenas to be able to move?

Problem 9-1

How do euglenas react to light and dark?

How do you think euglenas would react if they were placed in an environment with light and dark parts? Make a prediction and record it in your notebook. Here is a method to test your prediction.

Materials
the tube of euglenas that you looked at before
masking tape   paper punch
black construction paper

Gathering data
Figure 9.3 shows one simple method to test how euglenas react to light and dark. You could try a method of your own.

How does light affect euglenas?

FIGURE 9.3
Light comes through the hole in the paper. The rest of this environment for euglenas is dark. Where will most of the euglenas be?

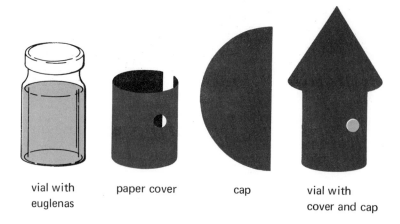

vial with euglenas   paper cover   cap   vial with cover and cap

The container or vial of euglenas should be filled almost to the top. Should you tighten the lid or stopper completely?

Take a piece of black paper large enough to wrap around the vial. Make a hole in it with the paper punch. Then tape the paper so it fits snugly around the vial. Cut and tape another piece of paper to form the cap, as shown in Figure 9.3. Light should come in only through the hole you made.

Now slip off the paper cover and stand it next to the vial. Draw a picture of the vial to show where the euglenas are before the experiment begins.

Slip the paper cover back on. Then put the vial where it will get bright light for two or three hours before the next class meeting. But don't put it in direct sunlight. Why not?

At the next class meeting, carefully remove the paper cover without stirring up the liquid. Stand the cover beside the vial. Sketch the vial to show where the euglenas are.

Recording data

Record your prediction. On the same page draw the euglenas before and after you covered them.

Analyzing data

Does your data fit your prediction? What evidence supports your answer? Did the euglenas react in the same way for your classmates?

Write a short paragraph describing how the euglenas' reaction to light might help them survive. Record your conclusions next to the experimental data.

Problem 9-2

Are other light conditions important?

Light and dark are not the only conditions euglenas experience in nature. Can you think of other light conditions that might influence euglena behavior? Discuss this question with your class.

You can investigate one or more new light conditions on your own. Make a hypothesis or a prediction about how euglenas will react. Then test it.

Materials
Your teacher will give you the materials and equipment you ask for.

Gathering data
Write down your plan of action. Include in it:

a. a statement of your hypothesis or prediction,
b. a list of the materials you need, and
c. a description of the procedures you plan to follow.

You may get some ideas from the setup in Problem 9-1. If you want to, you may change any part of your original plan. But be sure to write down any changes. Without an accurate description of an experiment, data does not have any clear meaning.

Ask your teacher or classmates if you need any help setting up your experiment.

Recording data
Record your observations below the written plan for your experiment. You may find it helpful to make drawings to illustrate your experiment.

Analyzing data
Does the data support your hypothesis or prediction? Did another group perform a similar investigation and get similar results?

You and your classmates can learn more about euglenas' light reactions by discussing each other's investigations. Afterwards, in a few sentences, describe the survival value of a euglena's reactions to light.

How does light affect euglenas?

Mastery Item 9-1

Where do euglenas live in this pond?

Many euglenas live in the pond in Figure 9.4. You wouldn't expect them to be evenly distributed throughout the pond. And they aren't. Where would they be on a bright, sunny day?

List the lettered parts of the pond, in order, from most euglenas to least. Then write a short statement explaining your predictions. *a.* is under some red food dye that was released into the pond from a nearby bakery; *b.* at the surface of an open area; *c.* under a boat; and *d.* near the surface of an open area that is partly shaded by trees.

FIGURE 9.4
Where would euglenas be in this pond on a bright sunny day?

How does light affect euglenas?

Key

d—This area provides most favorable light conditions, not too bright or dark. a—Red light due to the dye attracts euglenas, but not as much as shaded sunlight does. b—Direct sunlight causes most euglenas to move downward. c—Few euglenas would be attracted to the dark area under the boat.

Mastery Item 9-2

The sun and ocean-dwelling organisms

Organisms similar to euglenas live in the ocean. Scientists have found that these organisms make daily journeys between the ocean's surface and deeper waters. Figure 9.5 shows the vertical movement of one population of organisms during a 24 hour period. Study the diagram carefully. How might vertical movements benefit these organisms?

Key

The benefits you describe should be related to food production. Some light is necessary, but direct sunlight can be harmful. Discuss your answer with your teacher.

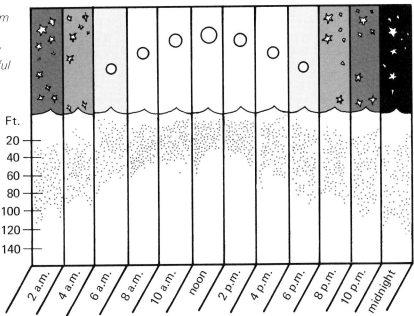

FIGURE 9.5

The organisms in this diagram move up and down in the ocean during each day. How could this movement be useful for the organisms?

FIGURE 10.1
Corn seeds can germinate even if they are still attached to the corncob.

FIGURE 10.2
A seed contains a young, inactive plant. The young plant and its food supply are wrapped in a protective coat.

Investigation

10 What influences seed germination?

What do seeds do? Seeds pass life on from generation to generation in many kinds of plants. Seeds are a way for a species to survive. Seeds are a way for life to almost stop . . . and then revive. They may **germinate** (JER-mih-nayt), or sprout, years after the parent plants have died.

Some germinating seeds are shown in Figures 10.1 and 10.2. Several kinds of **dormant** (DOR-munt), or inactive, seeds are shown in Figure 10.3.

FIGURE 10.3
Do these seeds show any signs of life?

How can seeds carry life from one generation of plants to another? How long do they keep this ability? The length of time that seeds can stay alive varies. Seeds from quaking aspen trees generally die after eight weeks in storage. On the other hand, red clover seeds have been germinated after 100 years.

Some seeds have carried life even longer. A Japanese botanist, Dr. Ichiro Ohga, obtained some water lily seeds collected from an old lake. The lake had filled with peat and had dried out. (Peat is ancient vegetation which has slowly rotted and packed solid.) These seeds were large, with a thick, hard coat around them. He was able to germinate them. An American chemist measured the age of some of the seeds. The chemist estimated that they were about 1,000 years old.

Some years ago a mining engineer found some Arctic lupine seeds under about 10 to 20 feet of frozen soil in northwestern Canada. A few years ago these seeds were

FIGURE 10.4
These Arctic lupines grew from seeds found in the Yukon. Botanists believe the seeds to be 10,000 years old. They look the same as Arctic lupines found today.

given to some Canadian botanists. They placed the best-looking seeds on wet blotting paper in a dish. Six of the seeds germinated within two days. The plants in Figure 10.4 grew from these seeds.

A rodent skull found with the seeds was estimated to be about 10,000 years old. Does this mean that lupine seeds can sprout after 10,000 years?

Dr. Frits Went, an American botanist, began an experiment in 1947 that is planned to continue for more than 300 years. Seeds from many kinds of wild plants in California were stored in separate, sealed tubes. These tubes are kept in an insulated room.

Ten years later some of these tubes were opened, and seeds of each type were tested for germination. About the same number of seeds germinated now as in 1947. Plans have been made to test some of the seeds for germination every 10 years until the year 2247.

While you were reading about these seeds, perhaps you asked yourself these questions:

How can seeds germinate after such a long time? Since they were found outdoors, why didn't the seeds germinate long ago? Can you make seeds germinate earlier or later that they would in nature? What is the purpose of Dr. Went's experiment?

In the Problems that follow, you will investigate dormant and germinating seeds. You will make seeds germinate, and test their responses to the environment. The data you gather may help you find some answers to these questions.

Problem 10-1

How do seeds respond to water?

As seeds ripen they dry out. They need to take in water before they can germinate. In this Problem you will investigate how water affects seeds. Does it affect the way seeds look? Can you look at a seed and tell when it is beginning to germinate? Does water affect the time it takes seeds to germinate?

What influences seed germination?

Materials

Your group will need these items:

2 Petri dishes or other containers for germinating seeds
8 disks cut from paper towels or pieces of filter paper
triangular file
40 untreated seeds of *one* kind: black locust, honey locust, or silk tree
20 acid-treated seeds (the same kind as the untreated seeds)
small plastic bag
metric ruler

Gathering data

a. Take four pieces of filter paper, or cut four disks from paper towels as shown in Figure 10.5a. (They should fit inside the smaller dish.) Label one as shown in Figure 10.5b.

b. Place them in the Petri dish with the labeled disk on top. Then add water as illustrated in Figure 10.5b and c.

c. Take 10 seeds from one container marked "untreated." Space them over Section 1 in the dish, as shown in Figure 10.5d.

d. Take 10 more untreated seeds and file a small groove in each one. File just through the seed coat on one flat side of each seed. Then put them in Section 2 with the grooved side down.

e. Space 10 acid-treated seeds evenly over Section 3. Your teacher soaked these seeds in concentrated sulfuric acid for 20 minutes before class. Then the acid was washed off with water, and the seeds were dried.

f. Cover the dish and place it in a small plastic bag. This will slow down evaporation.

g. Put your dish on a flat surface where your teacher tells you.

h. Add more water when necessary. Otherwise, you won't have to open the dish.

i. Record the numbers of seeds that germinate in each section each day.

How does water affect seeds before they actually sprout? Here is a way to tell without disturbing the seeds you just planted.

a. Prepare another Petri dish with filter paper or paper towels.

What influences seed germination?

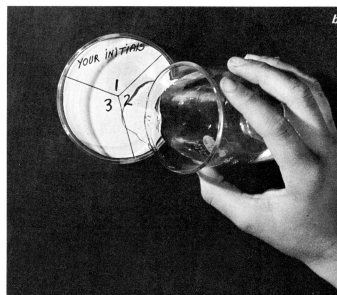
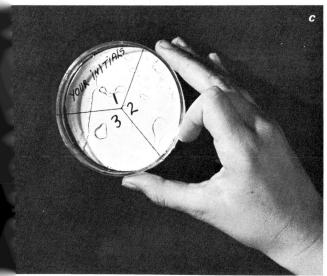
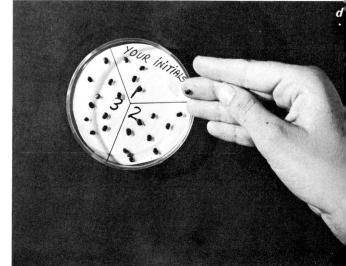

FIGURE 10.5
Preparing dishes to test how water affects germination: After you have placed the paper disks on the Petri dish, add water. Place 10 untreated seeds, 10 filed seeds, and 10 acid-treated seeds in the dish, as in d.

b. Before planting the same three kinds of seeds, measure them. Line up the 10 seeds in each group end to end. Measure this whole length in millimeters (the smallest unit on the metric ruler).
c. Space each group of seeds in the dish, just as you did before. Cover it.
d. Tomorrow and the next day remove one group of seeds at a time. Line each group up on a damp paper towel and measure the length. Then replace the seeds, cover the dish, and put it back.

Maybe you can think of a better way to observe changes in seeds before they germinate. If you can, use your method.

Recording data

You need two data tables. One is for the length of the seeds you measured. Also write down any other changes you notice. The second table is for germination data. Record daily the number of seeds that germinate in each section of the dish. Keep a record for seven days.

Analyzing data

Look at the data you got from measuring your seeds. Discuss this data with your group and other classmates who performed this experiment. Try to answer these questions.

a. Did any of the seeds change after they were kept in water for two days? What can you infer was happening?
b. Did you observe any differences among the three groups of seeds?
c. If some groups changed more than others, was only water responsible? What is your evidence?

Can you now predict which group of seeds will germinate first, second, and third? Record your predictions on the data table, and report your findings and predictions to the class.

At the end of your experiment, look at the data. Were your predictions confirmed? Discuss all the data with your group. You can also exchange results with other groups doing this experiment. Write a short report of your investigation for your classmates who worked on other Problems. The report should include a short statement of the questions or hypotheses you investigated. Also include your data, your analysis of the data, and any new questions or problems that come up.

What influences seed germination?

FIGURE 10.6
The newspaper wrapping helps this ear of corn stand up. Could the seeds at the top of the ear on the right sprout?

Here are some questions to help you analyze your data for your report.

a. What effect does acid or filing through seed coats have on germination?
b. What inferences can you make to explain the data from the untreated seed group?
c. What hypothesis can you make about the way other kinds of seeds might react to the same treatment?
d. What good is it for a seed to have a coat that resists letting water through?
e. Does this experiment help you answer any question from the beginning of this Investigation?

If you have thought of any other experiments to do, write down your plan of action. Check with your teacher before starting.

Problem 10-2

How does temperature affect germination?

If you did Problem 10-1, you studied the effects of water on germination. In this Problem you will experiment with temperature as the variable factor.

Can you remember from Investigation 7 how goldfish gill beats were affected by temperature? The goldfish reacted to temperature changes by breathing faster or slower. Will seeds germinate faster or slower depending on the temperature? Do different seeds germinate better at different temperatures?

Materials
Your group will need these items:

3 Petri dishes or other containers for germinating seeds
12 paper disks cut from paper towels or pieces of filter paper
2 kinds of seeds, 30 of each kind (You can choose the kinds from the list that your teacher provides.)
3 small plastic bags

Gathering data
a. Take 12 pieces of filter paper, or cut 12 disks from paper towels as shown in Figure 10.7a. Label the top ones as shown in Figure 10.7b.
b. Place four disks in each of three petri dishes with the labeled one on top. Then add water as shown in Figure 10.7b and c. Do you think it is important for all three dishes to have the same amount of water?
c. Space 10 seeds of one kind over half of a dish, as shown in Figure 10.7d. Add 10 seeds of another kind to the other half. Prepare your other two dishes in the same way.
d. After you have planted your seeds, cover the dishes. Place each dish in its own plastic bag. This will reduce evaporation, especially at the warmer temperatures.
e. Keep the seeds in a place where the temperatures will be nearly constant. One dish (50°F or 10°C) must be kept in a refrigerator. The second and third dishes (75° and 95°F, or 24° and 35°C) will probably be kept in incubators. (See the Appendix to change between °F and °C.)

What influences seed germination?

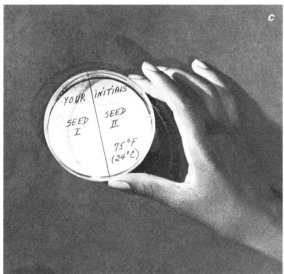
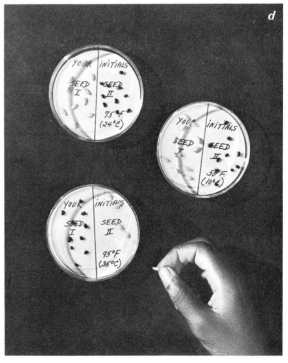

FIGURE 10.7
Preparing dishes to test how temperature affects germination

f. Check the dishes daily for germination. Add more water if necessary. Otherwise, do not remove the cover unless it fogs up and you can't make observations.
g. Keep a record of your data for seven days. Some seedlings may cover up seeds that haven't germinated yet. Remove the seedlings if necessary.

Recording data

In a data table, record for each day the *total* number of germinated seeds in each half of each dish. For example, take corn seeds incubated at 75°F (24°C). Suppose no seeds germinated during the first three days. Then six germinated on the fourth day, and two on the fifth. The numbers to record would be, 0, 0, 0, 6, and 8. Allow enough space in the table for seven days' data.

Write down below the data table any other observations you can make. These observations may help you to interpret the data on germination.

Analyzing data

Discuss all the data with your group. You can also exchange results with other groups who did this experiment. Write a short report of your investigation for your classmates who did not work with this Problem. The report should include a short statement of the questions or hypotheses you investigated. Also include your data, your analysis of the data, and any new questions or problems that came up.

Here are some questions to help you analyze your data for your report.

a. In what order did the groups of seeds begin to germinate?
b. Did any kind of seed germinate at the same rate at all three temperatures?
c. Could it be helpful for seeds to react to different temperatures?
d. Did some seeds fail to germinate because water did not get inside? Do you have data to support this inference? You can show these seeds to people who worked with Problem 10-1 and ask what they think. You could also cut open some of these seeds. How do they compare to sprouted seeds?

What influences seed germination?

e. Does this experiment help you answer any of the questions from the beginning of this Investigation?

If you have thought of any other experiments to do, write down your plan of action. Check with your teacher before starting.

Problem 10-3

Do seeds respond to light?

Green plants (those with chlorophyll) can't live very long without light. But what about their seeds?

Most seeds lose their green color as they ripen. Unlike their parent plants, they don't contain chlorophyll. Do seeds need light to stay alive? Do they need certain light conditions to germinate? In this Problem you will investigate these questions.

Materials

Your group will need the following items:

2 Petri dishes or other containers for germinating seeds
8 disks cut from paper towels or 8 pieces of filter paper

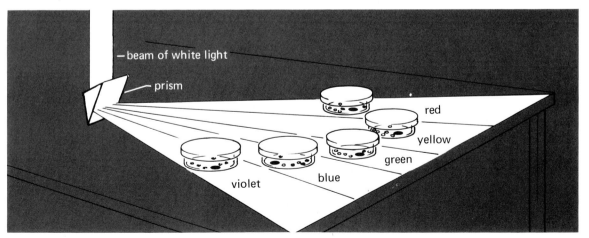

FIGURE 10.8
The prism breaks white light up into different colors. How do seeds respond to each color?

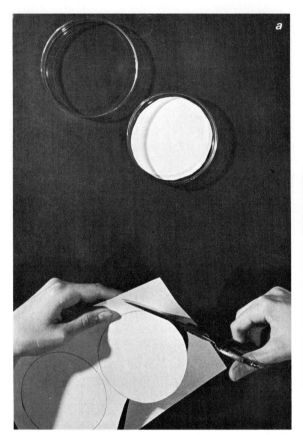
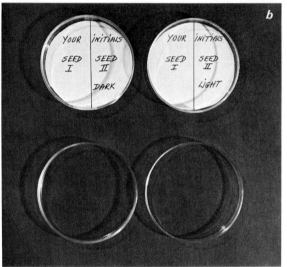

FIGURE 10.9
Preparing dishes to test how light affects germination

2 kinds of seeds, 20 of each kind (Choose one kind from each group that your teacher provides.)
black plastic sheeting (enough to completely enclose one Petri dish)
small plastic bag   scissors
tape   eye dropper

Gathering data

a. Take eight pieces of filter paper, or cut eight disks from paper towels as shown in Figure 10.9a. They should fit into the smaller dish. Label two of them as shown in Figure 10.9b.
b. Place four disks in each dish with the labeled disk on top.
c. On half the dry paper, space 10 seeds of one kind as shown in Figure 10.10a. Add 10 seeds of another kind to the other half. Prepare your second dish exactly the same way.

FIGURE 10.10

Planting seeds to test the effects of light

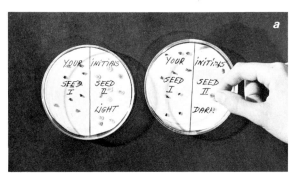

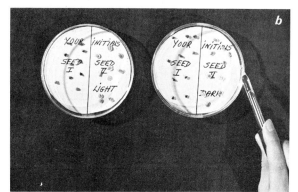

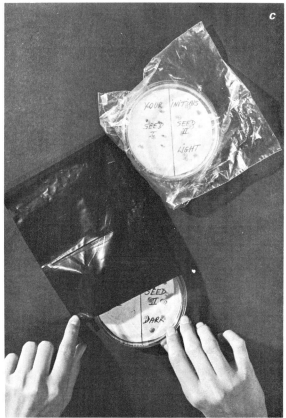

d. Now, quickly but carefully add water along the edge of the dish marked "dark." Do it as shown in Figure 10.10b. The disks should get good and wet. But if you add too much water the seeds will float around. Why should you add the same amount of water to each dish?

e. *Within a minute* after the water is added, seal the dish in the black plastic sheeting. See Figure 10.10c. *No light must enter.* Set the dish in the place your teacher has indicated.

f. Add water to the dish marked "light." Seal it in a clear plastic bag, also *within a minute* after adding water. These seeds should receive continuous light. Artificial light can be provided overnight. Do you think the temperatures of both dishes should be the same?

g. Don't open your dishes for the next seven days. You can observe the lighted seeds without opening the dish. Record how many seeds germinate in this dish each day. What else can you observe?

h. At the end of seven days, open both dishes for observation.

Recording data

Record how many seeds of each kind germinated in the light. Record how many seeds of each kind germinated in the dark.

Be sure to record other observations, especially any differences between the seedlings grown in the light and in the dark. These observations may help you explain the data on numbers of germinated seeds.

Analyzing data

Discuss all the data with your group. You can also exchange results with other groups who did this experiment. Write a short report of your investigation to present to your classmates who did not work with this Problem. The report should include a short statement of the questions or hypotheses you investigated. Also include your data, your analysis of the data, and any new questions or problems that came up.

Here are some questions to help you draw conclusions from your data.

a. Did any kind of seed germinate better in darkness? In light? If so, what inferences can you make?
b. What was the purpose of not observing seeds in the dark until the last day of the experiment?
c. You have observed the effect of total darkness and total light on seeds. Does your data tell you the best light conditions for germination?
d. Do you think some seeds might need both light and dark? How could you test this?
e. Were seedlings grown in the dark different from those in the light?
f. You have no data to show when the seeds in the dark germinated. Can you think of a way to get this data?

If you have thought of any other experiments to do, write down your plan of action. Check with your teacher before starting.

The class has probably learned a lot about seed germination. Now is the time to combine what different classmates

What influences seed germination?

FIGURE 10.11
What factors might affect the germination of these seeds in a field?

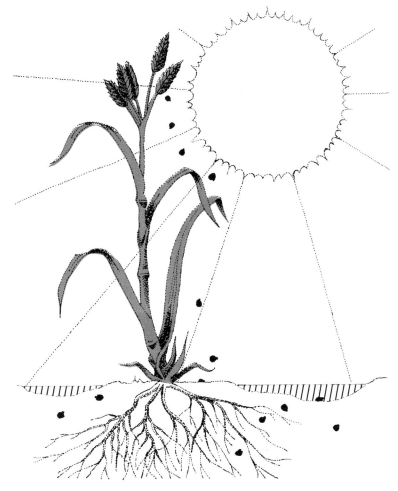

have learned. Then you will better understand germination and seed ecology.

Look at Figure 10.11. It shows a cut-away view of the soil in a field. Notice where the seeds are. What environmental factors could affect germination above and below ground? Assume the time of year is June.

How could germinating on top of the soil help seeds? How could germinating underground? Could it handicap seeds to germinate in each place? (See Figure 10.12.)

Seeds for many plants ripen and fall to the ground in August and September. Usually these seeds do not germinate until the next spring. What environmental factors might keep the seeds dormant all winter? What factors might make them germinate in the spring? How does seed behavior help the developing plants to survive?

FIGURE 10.12
Look at the root and sprout on the coconut. What conditions do coconuts need for germination? Do you think these conditions are found on this beach?

What influences seed germination?

Mastery Item 10-1

The best way to germinate seeds

Did Investigation 10 help you understand seed behavior, especially germination? Do you understand how water, temperature, and light affect germination? This Mastery Item should tell how well you understand.

Collect some new seeds from weeds or other plants outside. Or bring seeds from home, for example, from vegetables. These seeds should be fresh. They should have ripened during the last growing season. Older seeds may have lost their ability to germinate. If you cannot supply your own seeds, your teacher will give you some.

What environmental conditions do the seeds need to sprout? Your job is to find the *quickest* way to make them sprout. Write down how you will try to find out. Discuss your plan with your teacher before you begin.

Key

You have mastered this Item if you did these things:

a. Exposed your seeds to several temperatures, exposed them to light and darkness, and treated some seeds to help them absorb water.
b. Exposed at least 10 seeds to each condition.
c. Made daily observations and recorded them.
d. From your data, describe which conditions made the most seeds germinate in the least time.

Mastery Item 10-2

What makes peppergrass sprout?

Why do some seeds need to be at the surface of the ground to germinate? The following experiment was performed to find out.

Peppergrass is a common weed that produces small seeds. One hundred peppergrass seeds were scattered on moist soil in three trays. See Figure 10.13. The seeds in Tray 1 were not covered with soil. The seeds in Trays 2 and 3 were

152 What influences seed germination?

FIGURE 10.13

Why do peppergrass seeds germinate at the surface of the soil?

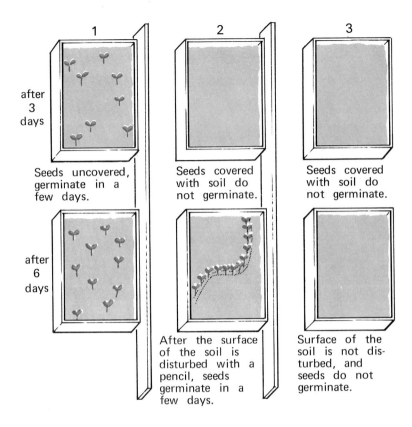

covered with one-fourth inch of moist soil. Clear plastic sheets were spread over all the trays to prevent drying. They were watered equally and kept in the same place at the same temperature.

In three days, many peppergrass seedlings sprang up in Tray 1. No seedlings appeared in the other two trays. Then a pencil was gently drawn through the soil in Tray 2. None of the uncovered seeds had yet begun to germinate. But three days later seedlings sprang up in Tray 2. Still no seedlings appeared in Tray 3.

Answer the following questions to see if you understand this experiment.

a. The seeds in all three trays would probably have germinated at the same time if they had: 1. better temperature, 2. more water, 3. better soil, 4. been planted on the surface. (Choose one condition.)
b. Which condition was probably changed most by drawing the pencil through the soil in Tray 2? 1. amount of water

for the seeds, 2. the seed temperature, 3. the amount of light that reached some seeds, 4. The seed coats were cracked.
c. Which conditions were kept the same for all seeds throughout the experiment? 1. light and air, 2. moisture and temperature, 3. moisture and a smooth soil surface, 4. moisture and the depth of the seeds in the soil.
d. Small seeds should be planted at the surface of the soil because: 1. they cannot grow up through much soil, or 2. certain conditions are better for germination there. What information from the experiment helped you choose an answer?
e. Tray 3 was not disturbed. Does it help you interpret the results of this experiment?

Key

a. (4) b. (3) c. (2)

d. (2) Seeds in Tray 2 did not begin to germinate until they were brought to the surface.

e. Tray 3 served as the control in the experiment. When no environmental factors were varied in Tray 3, no seeds germinated. Therefore, the differences in germination among all three trays were probably caused by the deliberately varied factor, light. That is a reasonable inference.

FIGURE 11.1
How does this plant get the water it needs?

Investigation

11 The water needs of plants

Plants, like animals, respond to changes in their environment. Plants are especially affected by the water supply. They can't live without water. Water carries food and gases to all parts of a plant. The amount of water plants take in and give off depends on the environment. When you have completed this Investigation, you should understand how environmental factors influence the water needs of plants and even influence plant shape.

Problem 11-1

Why do plants wilt?

Look at the plant the teacher has brought in. Why do you think it is droopy or wilted?

You have probably made the hypothesis that the plant lacks water. This plant was not wilted two days ago. Where did the water go? Did all of the water evaporate from the soil? Or did the plant use up the water? Develop a hypothesis with your group which you think explains this problem. You can test your hypothesis in the experiment below.

Materials

3 potted plants 2 rubber bands
2 plastic bags   balance

Gathering data

Your group will need three potted plants. Treat your plants according to the following directions.

a. Place a strip of masking tape on each pot for a label. Letter the pots A, B, and C as shown in Figure 11.2. Also mark each pot so you can tell it belongs to your group.
b. Do not treat plant A. See Figure 11.3a. Put a waterproof plastic bag around the pot of plant B. Fasten the top of the bag around the bottom of the plant stem, as shown in Figure 11.3b.
   Place plant C, pot and all, into a plastic bag. Fasten the plastic bag securely, as shown in Figure 11.3c.
c. Record the masses of all three plants.
d. Set all three plants in a bright part of the classroom, but not in direct sunlight. What questions does each pot represent? Write the questions in your notebook.
e. The next day reweigh the plants and record the masses.

Recording data

Make a chart on which you can record all the data: the dates and times that measurements were made, masses, and any other observations you think are important.

Compare the original masses with the measurements you just made. Are there any differences? If there are, record them for use in a class discussion.

FIGURE 11.2
Label each plant so you can easily keep track of it.

The water needs of plants   157

FIGURE 11.3
Why are three plants used in this experiment? What questions will each pot answer?

a

b

c

The water needs of plants

Analyzing data

The teacher will ask someone from each group to record any changes in mass for plants A, B, and C on the chalkboard. When this is done, discuss the data with your group. If there was a loss of mass, do you think the water was lost from the soil, from the plant, or both? Was your hypothesis about water loss correct?

Problem 11-2

What increases water loss in plants?

What environmental factors affect the amount of water a plant loses each day? What laboratory experiments could you set up to answer this question? A very simple apparatus called a **potometer** (po-TOM-uh-ter) measures how fast plants take up water. (See Figure 11.5 on page 160.) Plants lose water at about the same rate they take it up. So you can use the potometer to infer water losses.

In the first part of Problem 11-2 you will learn how to use this bubble potometer. Then your group can choose an environmental factor to investigate. You will test its effect on water loss in plants.

Materials
potometer   scissors
boiled water   bucket or plastic dishpan
leafy branches

Gathering data

The bubble potometer is an excellent tool for a curious person. It works something like the manometer you used to measure gas exchange in Investigation 6. The main difference is that with the potometer you follow the movement of an air bubble in a stream of water. In Investigation 6 you followed a drop of liquid in a column of air.

As a plant loses water from its leaves, it pulls more water into the leaves from the stem. That pulls up water from lower down in the stem and from the roots. (See Figure 11.4.) If a glass tube is substituted for the stem, you can watch the water moving. If you place a bubble in the glass tube, you can measure how fast the bubble moves.

FIGURE 11.4
Water enters a plant through the roots and escapes through the leaves. How fast does the water travel through the plant?

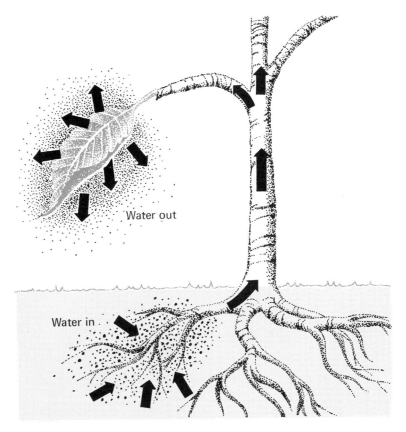

Follow the directions carefully. Your potometer should work well the first time you try it.

a. Join a piece of thin rubber (or plastic) tubing to glass tubing as shown in Figure 11.5. The glass tubing should be the kind with a very small hole in the center. It is called **capillary** (CAP-ill-air-ee) tubing. *Do not push hard on the glass tubing. It can break and cut your hand.* Wet the end of the glass and gently slide it into the rubber tubing.

b. Fill the sink or dishpan with water. Choose the plant stem to be used. *While the cut end of the stem is still underwater,* trim off about one inch of the stem. Do not bring the cut end of the stem out of the water at any time. Try not to wet the leaves of the plants. If you do, blot them dry with paper towels.

c. Now take the rubber tubing and capillary tubing you joined. Put them completely under the water in the same container. Squeeze the rubber tubing to get rid of all the air. Gently slide the rubber tubing over the cut end

FIGURE 11.5
The bubble potometer measures how fast plants take up water. What other information could you get from it?

of the stem. The tubing should fit tightly around the stem. (See Figure 11.5.) Do not tear the outside covering or bark of the stem. A piece of thin wire may be tightened around the tubing to insure an airtight fit. Leave the plant and the connected tubing in the water.

d. Set up a ring stand and container of *boiled water* that has been cooled to room temperature, as shown in Figure 11.5. Boiled water is used because boiling will drive out most of the air. Unwanted bubbles are less likely to form in the tubing.

You are now ready to move the plant and tubing to the ring stand. Place your finger over the open end of the capillary tubing. Gently move the plant and tubing to the ring stand. Place the end of the capillary tubing in the water in the beaker. Tighten the clamp over the tubing so the plant is held firmly.

e. Tape a ruler to the capillary tubing. The lower end of the ruler should be near the surface of the water. The ruler allows you to measure the distance the bubble travels.

f. There are two ways to get a bubble into the tubing. Either tip the container of water or raise the clamp holding the branch, until the end of the capillary tubing is just out of the water. Touch the end of the tube with a dry paper towel. It will absorb a drop of water and make an air bubble in the tubing. When you have replaced the tubing in the water, you are ready to begin.
g. Time the moving bubble with the second hand on your watch or on the wall clock. Record the time and distance traveled. Your data will be easier to interpret if you measure time in regular intervals such as 10 seconds, 30 seconds, or one minute. When the bubble has traveled almost to the rubber tubing, it should be stopped. Stop it by gently pinching the rubber tubing to push the bubble down the tube.
h. Take the average of several readings. That should give an accurate rate of water uptake under these conditions.

Now you know how to measure the rate of water loss from a plant.

You have data on how fast one plant takes in and loses water. What were the environmental conditions under which you gathered the data? For example, what was the temperature? The humidity? Was there a draft or breeze in the room? What were the light conditions? A sample chart for recording this kind of data is shown in Figure 11.6. Make

FIGURE 11.6
Make a chart to record your data for the experiment with the potometer. You may use a chart like this one, or make up one of your own.

Plant - Coleus	Humidity - 35%
Condition - healthy	Temperature - 75°F
Number of leaves - 16	Light - bright (150 foot-candles)

Variable (breeze)	Rate of water movement
slow	
medium	
fast	

a complete record of the conditions that influence your data in this experiment.

What will happen if you change one of these factors? Will it affect the rate at which the plant takes up and loses water? Talk with the other members of your group about which factor you would like to test. Write up your experimental design and discuss it with your teacher before you begin.

The following suggestions may help you set up your experiment. You can vary light intensity by moving a light source toward the potometer. For example, start with the light 2 meters away and move it 25 centimeters closer each time. Give the plant five minutes to adjust to each new intensity. (See Figure 11.7.) Record the distance of the light source or measure the amount of light with a photographic light meter. A fluorescent light is cooler than an ordinary bulb. It is less likely to overheat the plant.

You can change air movement from a light breeze to a strong wind by using an electric fan.

FIGURE 11.7
Change the distance between the light source and the potometer to see if light affects the rate of water intake.

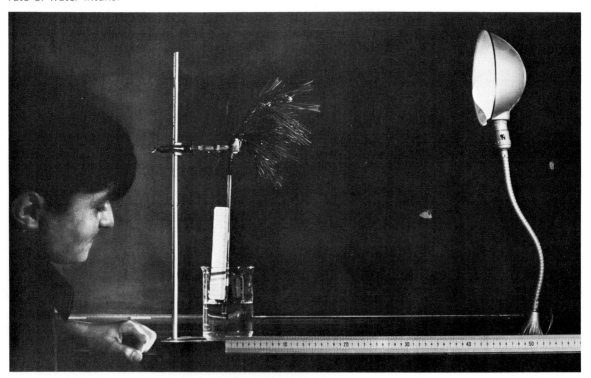

To change the temperature or humidity, cover the plant with a sheet of clear plastic. (See Figure 11.8.) For a humid environment you can put wet sponges or containers of water under the plastic. For a dry environment you can add chemicals that dry the air. Room radiators or heating tape are safe sources of heat.

Recording data

Keep a careful record of your data and your conclusions. You should record how changing each factor affects the uptake of water. Each group will report its experimental findings to the class.

FIGURE 11.8
Cover the plant with clear plastic. How can you test whether temperature and humidity affect the water intake?

Analyzing data

Talk with other groups about their data. Think of a way to put the group data on the chalkboard so everyone can understand it.

See if the class can agree on how each factor affected water loss. In what kind of environment would plants tend to lose water fastest? In what season in your area do you think plants lose the most water? The least? How could you find out?

Investigating problems usually raises new problems. You may want to set up another experiment to get data on a new question. Write out your hypothesis first. Then discuss your idea with your teacher.

Problem 11-3

Does leaf area affect water loss?

Does the number of leaves on a plant affect how fast it loses water? You can easily gather data to answer this question. All you need is your potometer and a plant. The plant can be a new one or an old one, depending on the supply of plants.

Materials
potometer   plants
boiled water

Gathering data

An easy way to answer this question is to measure water flow through a plant as you reduce the number of leaves (Figure 11.9). You already know how to set up the potometer, but read over the following hints before you start.

a. Once you cut off the leaves, you can't replace them. So plan your steps carefully.
b. Try to make measurements with at least five different leaf areas. You might use numbers such as 100 per cent of the leaves, 80 per cent, 60 per cent, 40 per cent, 20 per cent, and 0 per cent.
c. You might want to set up your experiment using factors that will give the largest water loss.

FIGURE 11.9
Does changing the number of leaves or needles affect how fast the plant loses water? Remember you can't replace the leaves you cut off.

Recording data

Keep a careful record of your data. When you are through, plot the data on a graph. Show the number of leaves on the vertical axis and the rate of water loss on the horizontal axis. (See the Appendix on graphing if you need help.)

Analyzing data

Compare your data with that of other groups. Can you make a general statement about how leaf area affects water uptake and water loss in plants? In what kind of an environment might fewer leaves be an advantage to a plant? Do you think the size of leaves affects water loss? Why?

Mastery Item 11-1

What are greenhouses for?

Many people use greenhouses to grow plants. Think of as many advantages as you can for growing plants in greenhouses. Why might it be easier than growing them outside?

Key

You should have said that greenhouses allow people to control important environmental factors such as temperature, humidity, air movement, and light.

FIGURE 11.10
Can you think of any advantages of growing plants in a greenhouse instead of outdoors?

The water needs of plants

Mastery Item 11-2

Adapting to climate

The shapes of plants have been partly determined by their environment. Looking at the shape of plants can help you infer where they live. Look at Figure 11.11. Which of these plants is more likely to live in an area of low rainfall? Explain your answer.

Key

The **cholla** (CHOY-uh) is a kind of cactus that grows in southwestern United States, where it is very dry. Instead of leaves it has spines. This reduces the water loss during dry times. The cholla cactus also has a thick stem for storing water.

FIGURE 11.11
Which of these plants is most likely to live in an area of low rainfall?
left, *cholla*
right, *tree fern*

FIGURE 12.1
These are three different types of sowbugs: Oniscus, Armadillidium, and Porcellio.

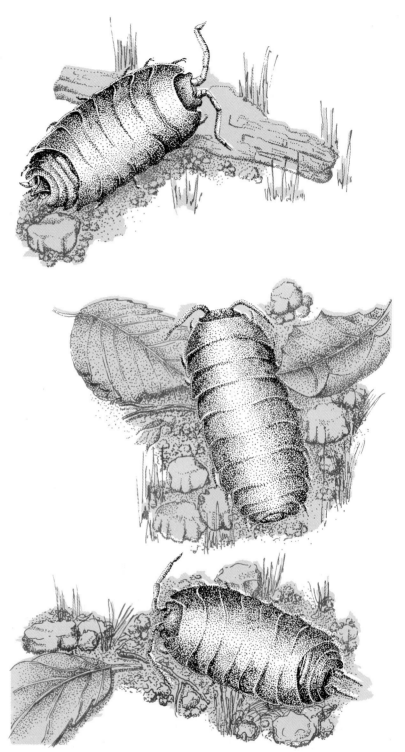

Investigation

12 What environment suits a sow bug?

FIGURE 12.2
You can trap sow bugs with a hollow potato.

What environment suits a sow bug?

In this Investigation you will be working with sow bugs. Sow bugs are also called pill bugs and wood lice. These animals can be found under rocks, boards, and logs almost everywhere. In cities they often live in basements and under boxes and rubbish. However, they are hard to find in winter or in hot, dry weather. Your teacher may have asked you to find some and bring them to class for this Investigation.

You can use the common sow bug to observe how the environment affects behavior. Your teacher will give you a container with several sow bugs, like the one in Figure 12.3.

Make observations to see what sow bugs are like. You can attract them to the surface by placing a damp paper towel on top of the soil. Feel free to handle them. How many sections does a sow bug's body have? How many legs does a sow bug have? The sow bug's gills are on its underside, but inside its shell. As in goldfish, the gills are for breathing. They must be moist to work.

Do sow bugs have eyes? Antennae? Look at the sow bugs with a hand lens.

Move the sow bugs from one spot to another and observe their reactions. You can put a tiny dot of paint on the back

FIGURE 12.3
This "sow bug environment" can hold several sow bugs. What are some of the environmental factors where sow bugs live?

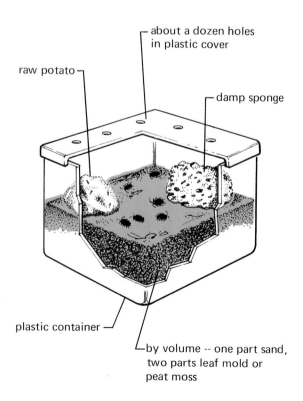

of a sow bug as its identification mark. Then you can keep track of individuals throughout this Investigation.

Take another look at the sow bug container. This time try to identify *specific factors* of the environment in which these sow bugs are living. Use all your senses in observing. Make a list of the factors. You can start with light.

Now, consider the following questions: Where were the sow bugs in the container when you first looked in it? Were they moving? How did they respond when you moved them around? When sow bugs have a "choice" of different conditions, which ones do they "prefer?"

What environmental conditions are most likely to cause sow bugs to change their behavior? Discuss these questions with your classmates. Have you developed some hypotheses about sow bug behavior that you want to test? Can you predict how light, temperature, and moisture might affect sow bug behavior? If you need help, Problem 12-1 may give you some ideas.

Problem 12-1

What makes a sow bug react?

Design a controlled experiment to test each of your hypotheses about sow bug behavior. You can do as many experiments as you have time for.

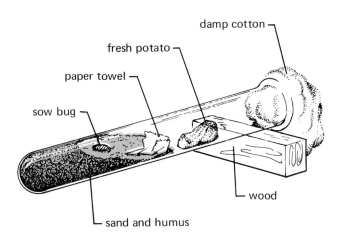

FIGURE 12.4

Container for an individual sow bug

 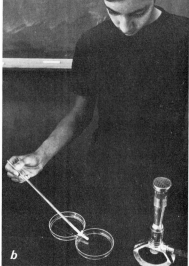 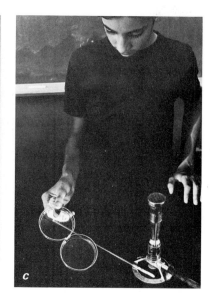

FIGURE 12.5
Making "choice chambers"
a. Heat the end of a glass rod.
b. Join plastic Petri dishes by melting their sides together with the hot glass rod.
c. As many as four dishes may be joined to a central dish. Sow bugs can easily move from one dish to another.

FIGURE 12.6
You can use this equipment to design experiments where a single factor varies. The dishes on the left were joined as shown in Figure 12.5. Plastic Petri dishes can also be joined by glass or plastic tubing. Make holes for the tubes with a hot glass or metal rod. When dishes are joined by tubes, the distances between dishes can be varied.

What environment suits a sow bug?

Materials

several sow bugs
various materials your teacher has provided (*Some of them are shown in Figures 12.4–12.8.*)

Gathering data

Before you actually begin your experiment, write down your hypothesis, your plan of action, and the list of materials you will need. Then discuss your plan with your teacher. Other people in your group may want to test different hypotheses. If there are enough sow bugs and materials, each person can work individually. Some of your classmates may be doing experiments like yours. Try to exchange ideas.

FIGURE 12.7
An environment where the humidity gradually changes: Put wet cotton at one end and drying chemicals at the other end. Wait about half an hour. Then put some sow bugs in. Use the test paper provided to measure humidity in different parts of the environment.

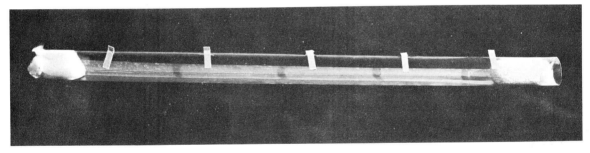

FIGURE 12.8
An environment where the temperature gradually changes: Heat one end of the chamber and cool the other end with ice cubes. Then put some sow bugs in. This apparatus should be used in dim light.

A final suggestion: Sow bugs get tired when you make them work hard. Be sure to let them rest frequently.

Recording data

The way you organize the data depends on the kind of data you get. For example, if you measure the time sow bugs spend in specific areas, make a data table. Later, you can make a bar graph of this data. If you need help organizing your data, look back at the data tables and graphs you used in Unit 1.

Analyzing data

Does your data support your hypothesis? Did any classmates perform the same experiment and get similar results?

What else have other classmates learned about sow bug behavior? You can discuss the findings each group presents to the class. Try to put together a general picture of sow bugs' reactions to environmental change. What conditions do they "prefer"? What conditions do they avoid? In what ways do they respond? In your record for this Investigation you ought to include a summary of this discussion.

Mastery Item 12-1

Where will the blowflies settle?

Blowflies are extremely common insects. You have probably seen them. Most of them are about the size of a house fly, and many of them are a metallic blue or green color. Sow bugs and blowflies react to temperature differences. But do you suppose they react in the same way?

The small table in Figure 12.9 has a metal top. There is a gas burner under the center of the table. There are ice cubes in the trough around the edges of the plastic top. About 30 minutes after the ice cubes are placed in the trough and the burner is lit, 25 blowflies are put in through the plastic cover. The cover prevents the flies from escaping.

When the flies become tired of flying, they will land somewhere on the table top. Your task is to predict where the flies will finally settle.

FIGURE 12.9

How do blowflies respond to temperature?

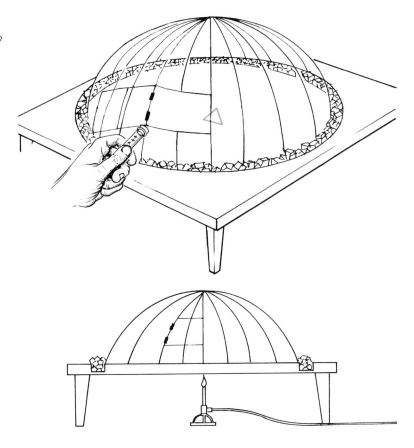

On a separate sheet of paper draw a square with the symbol △ in the center to represent the table top. Now, use X's to represent the individual flies, and mark on the paper where you think they will finally settle down. Then explain in a sentence or two why you put the X's where you did.

Key

The X's should be arranged in a roughly circular pattern somewhere between the symbol △ and the trough of ice cubes. The flies probably settled between the center and trough because the temperature there was most favorable to them. It was probably too hot at the center and too cold at the edges of the table. There are other ways to explain your arrangement of X's, especially if you arranged them differently. You can decide with your teacher how good your reasons were for making your pattern of X's.

Mastery Item 12-2

The survival value of animal behavior

Do you understand how an animal's behavior may help it to survive? You can find out by doing this Mastery Item.

Obtain several mealworms from your teacher. Observe and handle them to become familiar with their structure. Then try to discover some of the conditions that affect their behavior. If you need any materials or equipment, ask your teacher for it.

FIGURE 12.10
Observing mealworm behavior

What environment suits a sow bug?

Where in nature do you think the mealworms could live? Describe a place, based on the data you just gathered. Also, describe at least three environmental conditions that the mealworms would need to react to in order to survive in this place. Again, base your answers on data from your work with the mealworms.

Key

The place and conditions you described are acceptable if you have data to support your choices. Your data can be behavior and other characteristics of the mealworms that you observed. You and your teacher should discuss your answers to decide whether they are acceptable.

FIGURE 13.1
Sunlight doesn't reach down to the deep waters in this lake. What might this part of the lake look like if it received light?

FIGURE 13.2
Sunlight reaches all parts of the pond. What differences can you observe between this pond and the water in Figure 13.1?

Investigation

13 The energy needs of a community

What light conditions in Figures 13.1 and 13.2 would favor many kinds of organisms? Remember what you learned about euglenas. Which environment would they prefer? Would you expect to find animals that eat euglenas in both environments?

Earlier you discovered how euglenas react to light and dark. You studied only euglenas then, but in nature organisms never live alone. They are always interacting with each other. The animals, plants, bacteria, and other organisms that live in a place form a **biological community.** The pond in Figure 13.2 is an example of such a community.

In this Investigation you will put different pond water organisms together in jars to make small aquatic communities. Then you will see what happens when the communities live under different light conditions.

Problem 13-1

A community in the dark

In nature it is difficult to investigate how light affects a community. Other factors, such as temperature, wind, and minerals, also affect communities. In a laboratory, however, you can control these factors more easily. It is also easier to vary the factor you wish to study.

The aquatic communities you set up will contain euglenas and other tiny organisms that live in most ponds. Some of these communities will be kept in light, some in the dark. Several times during the next two weeks, you will observe the communities to see how each changes.

Materials

3 pint jars with paramecia and similar organisms growing in "hay soup"
euglenas and several other organisms
Pasteur pipettes or pieces of glass tubing with rubber bulbs on the end
microscope, slides, and cover slips
3 rubber bands   clear plastic sheets

Gathering data

a. Get three jars of "hay soup." Number the jars and label them so you know which are yours. Your teacher made the "hay soup" about a week ago by boiling some hay in water. Then the teacher added **paramecia** (*pahr*-uh-ME-she-uh) and similar organisms. (You will see what these organisms are like in a while.)

b. Add equal amounts of the following to each jar:
10–20 ml of euglena "soup"
10–20 ml of **Chlorella** (kluh-RELL-uh) or another kind of **algae** (AL-gee)

c. Set the jars aside and in your notebook make a data table like the one in Figure 13.3.

Now you are ready to observe the organisms in the jars. Each jar is like a small part of a pond community. Since you made all the communities alike, you can examine a sample from any jar. Later you can compare data with classmates who sampled other jars.

What part of a jar should you take a sample from? It will be important to get an accurate idea of the kinds and numbers of organisms. Each part of the jar may not contain the same organisms. You and your group should decide on the best way to sample the jars. Once you have decided, begin.

d. Take a sample from one of the jars, using a Pasteur pipette or glass tubing with a rubber bulb on the end.

e. Immediately prepare a wet mount on a slide. Use one drop of liquid.

FIGURE 13.3
Make a record of your data on aquatic communities. You can show the number of each kind of organism this way.
- **0** = none seen;
- **+** = very few;
- **+ +** = commonly seen;
- **+ + +** = many seen.

You may use a different system if you want. You may also classify the organisms in your own way.

		Date		
	Organisms			
Jar 1 light for 2 weeks	euglenas ciliates rotifers algae			
other observations:				
Jar 2 dark for 2 weeks	euglenas ciliates			

f. Focus your microscope on the liquid. Use 100X magnification.

Forget the Investigation for a while and just look at the shapes and motions of this microscopic world!

Some of the organisms may zoom so quickly through the water that you can hardly see them. You can slow them down with **methyl cellulose** (METH-uhl SELL-you-lohs). This is a gooey substance that won't hurt the organisms. Make another wet mount and add a tiny drop of methyl cellulose to the sample.

g. Scan the entire area under the cover slip. Figure 13.4 will help you identify the different kinds of organisms you see. You might have to increase the magnification to observe some of them clearly.

h. On your data sheet, record the kinds and numbers of organisms you see. Exchange data with your classmates so that you have a record of each community.

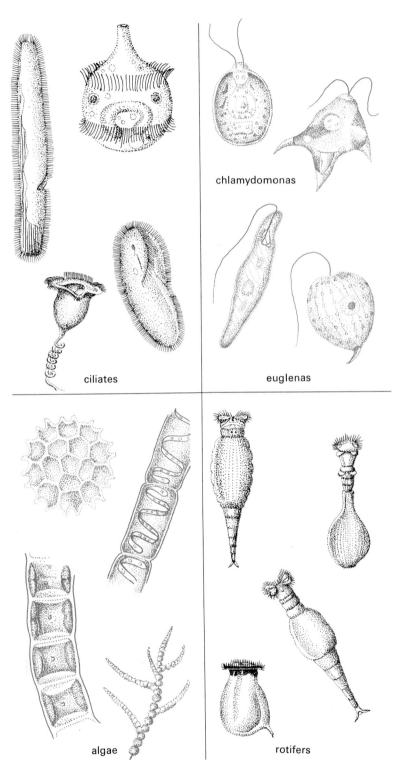

FIGURE 13.4
Some microscopic organisms that live in fresh water

Energy needs

This completes the first day's observations. Cover the jars with clear plastic. Should you punch holes in the plastic? Put Jars 2 and 3 in a dark box for the next week. Jar 1 should be left in the light. Figure 13.5 shows a possible experimental setup.

i. One week after the first sampling, sample each jar again. How will you sample your communities this time? Will one sample from each jar be enough? If you and your group have difficulty deciding on a sampling method, try several methods.

j. Return Jar 3 to the box. Keep both Jars 1 and 2 in the light for another week.

k. One week after the second sampling, sample the jars again. This is the last sampling. Record your data. If you are interested, you can continue making weekly observations on your own.

FIGURE 13.5
This setup shows one way to investigate how light affects your communities. Which jar is a control?

Your data will show how light can affect an entire community. But you won't know *why* the communities changed. Now you have to learn about the organisms you observed. You also need an idea of where light energy goes after it is trapped by algae and euglenas. The following information should help explain what happens in your communities.

Euglenas are already familiar to you from Investigation 9. You may also see some similar, green, fast-moving organisms called **Chlamydomonas** (Clam-ee-duh-MOE-nuhs). Euglenas usually reproduce by dividing lengthwise into two new individual cells. Under good conditions several generations can be produced in one day. Some of euglenas' relatives are made of several cells.

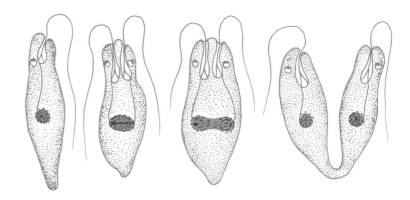

FIGURE 13.6
Euglena reproduce by splitting in the middle into two separate organisms.

Paramecia and their relatives are called **ciliates** (SIL-ee-ayts). Each ciliate is a single cell. They have tiny hairs on their bodies called **cilia** (SIL-ee-uh). The cilia wave around to move the paramecia through the water. They also sweep food into the paramecia's mouths. Ciliates feed mainly on organisms smaller than themselves, like bacteria. They reproduce by dividing into two new individuals. One or more generations can be produced daily.

Rotifers (ROH-tih-fers) are small animals with bunches of cilia on their heads. These cilia help rotifers move around and sweep food into their mouths. Unlike ciliates, rotifers are made up of many cells. They eat dead matter and tiny organisms, especially algae. In turn they are eaten by larger animals. Young rotifers hatch from eggs laid by adult females. Some kinds of rotifers can live for three weeks.

Algae (AL-jee) are tiny organisms containing chlorophyll. Usually they are green, but some look brown. Algae carry

FIGURE 13.7
Paramecia reproduce like euglenas. Each paramecia becomes two new ones. Several generations of paramecia can be produced in one day.

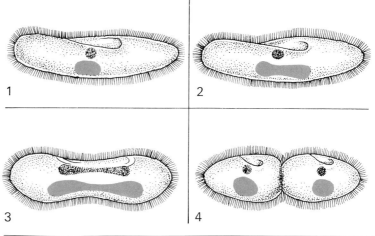

FIGURE 13.8
A rotifer eating algae

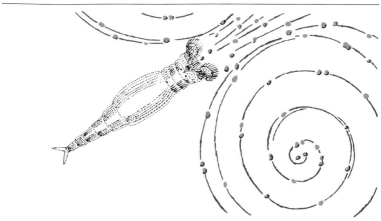

FIGURE 13.9
Some algae reproduce by forming a case, which opens to release several algae. Other algae reproduce by dividing to form two new algae.

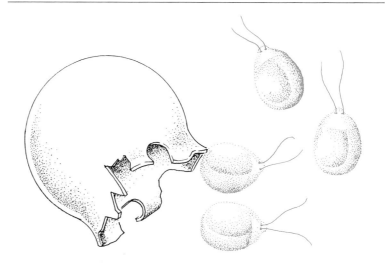

on photosynthesis to make food. The algae in the jars may have only one cell, or they may have many cells joined in a string. Most kinds can't move around. Algae provide food for many organisms.

Algae reproduce in many ways. Single-celled algae usually divide to make more single-celled algae. Algae that look like string can reproduce by adding cells at the ends of the string.

In a whole pond community you would also observe living things like those in Figure 13.10. This illustration shows another kind of relationship between the organisms in a community. Each kind of organism is a link in a **food chain.** Energy in the form of food passes from one kind of living thing to another along the chain. What is the source of energy here? What happens to most of the energy at any link?

Usually food chains branch out. For example, most fish eat many kinds of organisms. Branched and entangled food chains are called **food webs.**

Consider what you already know about how light affects plants and animals. Also think how the different organisms in your communities get food. Can you predict what will happen to your communities in two weeks? Write down your predictions on the page with your data table.

Recording data

Most of your data will be recorded in the table you made. If you are interested, you can sketch any organism you've seen that is not shown in Figure 13.4. Then you can try to identify these organisms in additional books that your teacher may provide.

Analyzing data

What effect does light seem to have on a community? Look at all the data your group has. Can you give a general interpretation of your data? If you have difficulty, use these questions as guides.

a. Which jar had the most living things? The least?
b. What happened to the numbers of organisms in Jar 2 after it was returned to the light?
c. What were the sources of energy in your communities?

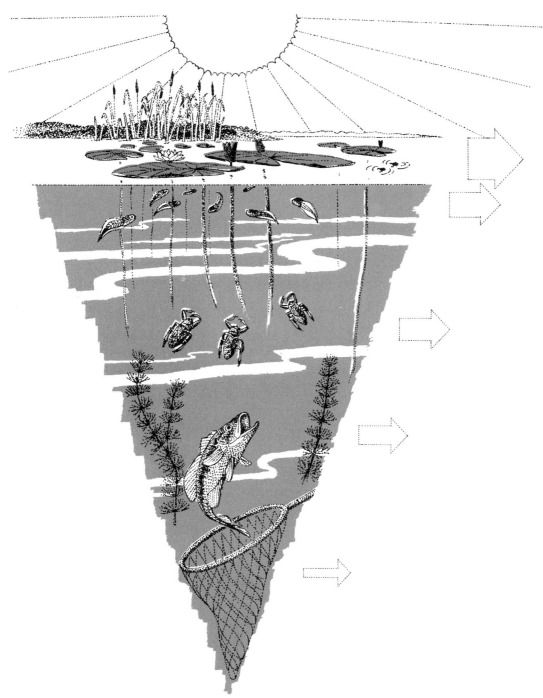

FIGURE 13.10
In this pond food chain the arrows at the right indicate energy used for life processes or given off when the organism decomposes.

188   Energy needs

Discuss your findings with the rest of the class. Then describe some possible food chains or food webs in your communities.

As your class discusses the data, you should come to understand why light is important to an aquatic community. But what about other biological communities? How is light important to grasslands, forests, and deserts? Do animals that come out only at night, like owls, need light?

Mastery Item 13-1

Life in a pond

Look at Figure 13.11. It shows some of the organisms in a pond community. What pathways does energy take through this community? In a paragraph or a drawing, describe the food web in this pond. The organisms shown in this pond community include: *a.* pickerel weed; *b.* waterlilies; *c.* elodea; *d.* water beetle; *e.* mosquito; *f.* turtle; *g.* frog; *h.* minnows; *i.* perch; *j.* sunfish; and *k.* bass.

Key

Here is a food web you may have described. You might have had other ideas, too. Discuss your answer with the teacher.

The energy from the sun is absorbed by the plants. Then, judging from the picture, beetles and minnows eat plants. Then they are eaten by turtles and larger fish. The fish may be eaten by man. Fish and frogs also eat insects like mosquitoes. The mosquitoes may get their energy from biting people.

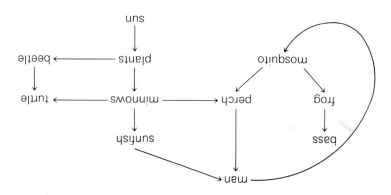

FIGURE 13.11
A pond community consists of many different kinds of life.

Mastery Item 13-2

Life on a coral reef

Coral reefs are walls of limestone that rise up near the surface of warm, shallow ocean water. The limestone is made of the skeletons of many animals called corals. These animals spend their whole lives in the same spot. Figure 13.12 shows how a coral reef may surround a lagoon of still water. The organisms that usually live on and around a reef are also shown.

FIGURE 13.12
A coral reef is a wall of limestone that rises up near the surface of warm, shallow ocean water. A coral reef is shown from the air (top), *and from underwater* (bottom).

Energy needs

The beginning of the food web is microscopic floating plants. Tiny animals and some small fish feed on these and other plants. The sponges eat tiny organisms and dead matter that they filter from the water. So do the smaller corals.

Anemones eat fish and other animals that they catch with their tentacles. Anemones don't move around much during their lives. Tuna, sharks, and other fish feed mostly on the smaller fish.

Suppose man dumps or spills thick oil and it washes inside a coral reef. What will happen to the populations of each organism described above? Write your predictions and reasons for making them.

Key

The thick floating oil will block out some, if not most, of the light. The plants will absorb less light energy and produce less food. Their populations will become smaller. So will the populations of tiny animals, sponges, and corals that eat the plants or eat animals that do. The anemones and larger fish will have fewer little fish to eat. Some of the animals that can't move may die off altogether. The populations of fish may die or move out of the area. However, they may not be able to find other similar places where they can live.

Unit three

Living things affect each other

Have you ever had a pet animal? Did you change its behavior to train it or to protect it? Did your pet learn how to change your behavior?

Your home, your school, your friends in the neighborhood—all would be somewhat different if you weren't there. Perhaps you would be somewhat different, too, if you lived in another neighborhood and had different friends.

In a forest each tree has some effect on the surrounding trees, the grass, and the animals that live there. All living things affect each other.

Here is a last example. Think of how you affect living things (including other people) if you like to drink milk, eat eggs, wear woolen sweaters, or play in parks. In fact, what do you do that doesn't affect other living things?

Unit Three will help you learn how to study the interactions between living things. You will also learn to:

a. Predict what will happen if the numbers of plants, animals, or microbes in a particular environment are changed.
b. Predict some of the ways the earth will be changed by a "population explosion."
c. Suggest how to use some of the land in your own community.

FIGURE 14.1
The coyote is a wild animal that looks like a dog. Do you think it is a harmful or a helpful animal?

Investigation

14 Coyotes and their prey

Nothing ever stays the same. The conditions of an environment, like temperature and light, are constantly changing. In turn, they cause changes in populations of organisms. Can you remember the changes in the aquatic populations (Investigation 13) that resulted from different light conditions?

Populations are affected by more than environmental conditions. Populations affect each other. Farmers and ranchers change natural environments. They do so to supply you with the foods you eat.

In this Investigation you will learn some of the problems these people create as they change the environment. Then you will be asked to propose what you think is the best solution to the problems.

Problem 14-1

The coyote problem

One perception of coyotes

The **coyote** (kye-OH-tee) is a kind of wild dog. It looks like a yellowish-grey collie. An average full-grown male weighs about 30 pounds. Females are a little smaller. Usually coyotes hunt, eat, and play at night and then sleep during the day. They live in dens in the ground in hidden places.

The map in Figure 14.2 shows that coyotes can be found from Alaska to Mexico. They are very **adaptable** (a-DAP-ta-bul). This means that coyotes can change their habits to fit into many environments.

The coyote lived in the grasslands of Kansas for many centuries. During this time, the environment probably did not change much. Then homesteaders arrived and caused drastic changes in the coyote's natural community. The homesteaders brought new kinds of animals. They plowed and fenced the range, and they hunted coyotes. The coyote had to move, or adapt to the new conditions, or die out. Many other animals have become rare since the white man came. Some of these, like the coyote, are meat-eaters. But the coyote adapted to man-made changes.

Man and coyotes have caused problems for each other. Farmers and ranchers claim that coyotes kill many of their animals. Therefore, they want to **exterminate** (ex-TER-min-ayt) or wipe out the coyotes. In 1949, the Agricultural

FIGURE 14.2
The shaded areas show where coyotes are found in the United States. How many different environments can coyotes live in? Are there any coyotes where you live?

Experiment Station in Kansas asked farmers and ranchers how much damage coyotes did in a year. Here are the answers they received.

Farmers estimated that coyotes killed $575,000 worth of chickens. Losses were greatest when farmers let tall weeds and crops grow near the chicken yards. Coyotes hid in the tall plants and captured chickens that wandered by.

Ranchers said they lost more than 6,000 calves, worth $325,000. Ranchers were sure that coyotes had killed some newborn calves. That happened when cows were left on the range to give birth. More often, ranchers found dead calves partially eaten by coyotes (they inferred).

Sheep ranchers said they lost almost $300,000 worth of animals in 1949. Most animals were lost in the spring. Lambs are small then, and coyotes need to provide food for their pups. Unguarded flocks were most often raided.

Farmers also estimated that coyotes killed $150,000 in turkeys, pigs, and other livestock. All these losses made farmers and ranchers spend extra money on guards, pens, and fences to protect their animals from coyotes.

A proposed solution

The farmers and ranchers wanted to exterminate the coyotes. They asked the state legislature to pass a law. It called on the state to do these things:

a. Increase the bounty, or reward, for killing a coyote from $2 to $5.
b. Try to kill all coyotes in livestock- and poultry-raising areas. Specialists in extermination would come from the Fish and Wildlife Service of the federal government. They would put an extremely powerful poison in meat and leave it where coyotes would find it.
c. Open all government lands to year-round hunting of coyotes.

What do you think about this proposed solution? Do you know enough about coyotes to decide whether this is a good law? Discuss these questions with your group.

Decide if you recommend passing the law. If you feel that you need more information to decide, go on to the next part. If you want to vote on a solution, turn to the bottom of page 203.

FIGURE 14.3
One view of coyotes: Farmers report huge losses of domestic animals because of coyotes.

Coyotes and their prey

A study of coyote food habits

For 10 years biologists at the Kansas Agricultural Experiment Station studied what coyotes eat. It is difficult to observe coyote behavior. Remember, they hunt and eat mostly at night. The biologists decided to analyze the food in the stomachs of dead coyotes. Hunters kill many coyotes each year for bounty and sport. The biologists asked hunters to bring them these coyotes. Figure 14.4 shows what the biologists found in 1,600 coyotes.

Notice that the bands on the graph swell and shrink. This shows how the coyote's food changes from one season to another. Coyotes eat what is plentiful. The supply of each kind of food changes during the year.

The data shows that almost two-thirds of the coyote's diet is rabbits and rodents. These are pest animals to farmers and ranchers. They are difficult and expensive to control.

Carrion (CAYR-ee-un) in Figure 14.4 is rotting meat. It comes from animals, including livestock, which have been

FIGURE 14.4

What coyotes in Kansas eat during the year: "Rodents" includes rats, mice, and other small wild animals. "Chicken" includes all poultry, whether freshly killed by coyotes or picked up dead. "All other" includes fruits, grains, insects, and snakes. What per cent of the coyote's diet is rodents?

dead for some time. Usually coyotes find the dead livestock on the range. They will also eat dead farm animals which farmers leave lying around. This way coyotes may develop a taste for farm animals.

What is the basic food of coyotes? Does it look like coyotes kill and eat lots of livestock? What time of year is the coyote the worst pest to ranchers and farmers? Do you think a coyote is being a pest when it eats carrion? Might coyotes actually be helpful to farmers?

The biologists found there are a few "killer" coyotes that eat almost nothing but livestock. One killer may cause most of the livestock deaths in his territory.

The study also showed that farmers and ranchers often blamed coyotes for damage done by dog packs, raccoons, and bobcats. Coyotes probably killed less than half of the sheep and calves for which ranchers blamed them.

FIGURE 14.5
Another view of coyotes: They eat what is most plentiful. Their diets change from month to month. A large part of their diet is rabbits.

Are coyotes valuable?

Do coyotes eat enough rodents and rabbits to make up for the livestock and poultry they kill? The biologists tried to answer this question. They used the idea of food webs to help them.

The food web in Figure 14.6 shows some animals that live together in Kansas. The arrows show who eats whom. The animals on the same level eat the same kind of food. The food web shows that rodents, rabbits, sheep, and cattle eat plants. You can infer that all these animals compete for grass. Because coyotes kill rabbits and rodents, more grass is available to cattle. How helpful are coyotes then? Remember that rabbits and rodents are their basic food. Figure 14.7 shows the value of coyotes as pest killers.

FIGURE 14.6
A food web in Kansas: Producers are plants. They produce energy from sunlight, therefore providing energy for all the other organisms in a food web. Animals that eat plants are sometimes called first-order consumers. Second-order consumers are meat-eaters; they eat the first-order consumers. Third-order consumers eat meat-eaters. They may also eat plant-eating animals.

1. Coyotes eat 3 million rabbits and 21 million mice each year in the state of Kansas.

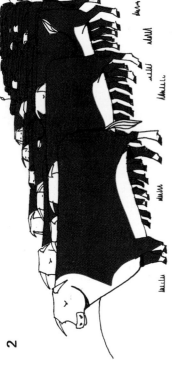

2. The amount of grass eaten by 3 million rabbits and 21 million mice in one year would, if not eaten by the rabbits and mice, feed 47,820 cattle.

3. It costs $18.00 per year to graze one steer in an open pasture.

4. Therefore, coyotes provide a service to Kansas farmers that is worth about $900,000 (the value of the grass that coyotes save from rabbits and rodents).

FIGURE 14.7
Coyotes can often be useful animals to have around. Does their value outweigh the harm they sometimes do?

Another look at the proposed solution

How do you perceive the coyote problem now? Would you vote for the law now? Would you change the law?

Discuss these questions with your group. If you want to vote on a solution, skip to the bottom of this page. If you are still uncertain about how to solve the problem, go on to the next part.

What controls coyote population size?

If people didn't hunt coyotes, would there be a "population explosion"? An environment has only enough food and shelter for a certain number of animals. A male and female coyote seem to need about one square mile of territory to hunt in. They will fight to keep other coyotes off their territory. When coyotes are crowded, fighting and disease kill some of the extra animals. Some die of starvation.

Food supply seems to be the most important control on the coyote population. When there are many rodents in the winter, more females breed. They also have many pups in each litter. The graph in Figure 14.8 shows how food and reproduction are related. In what years were the most rodents available? Was reproduction high in those years? In what years could the coyote have been a big pest to farmers? If you have trouble answering any of these questions, ask your teacher for help.

How would you solve the coyote problem?

Now you can vote for or against the proposed law, or not vote at all. Or you can make changes in the law before voting.

Discuss the following list of possible solutions with your group. Then choose by a vote *one* of the solutions listed below. (Or if you think of your own solution, write it down.) Predict what would happen to the coyote problem as a result of your decision. Also write down the reasons for your choice. If the groups disagree on what to do about coyotes, they can debate the question.

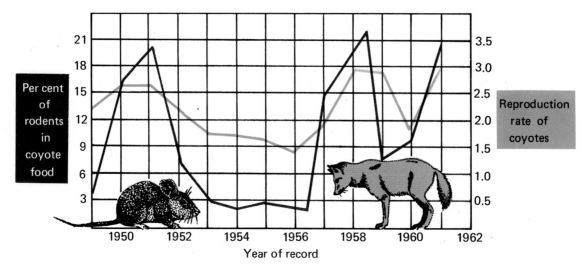

FIGURE 14.8
How food supply affects coyote reproduction. The vertical axis on the left shows what per cent of coyote food was rodents. This data was collected in the winter, before the breeding season. The vertical axis on the right shows the reproduction rate. It is high when many females breed and have large litters. It is low when few females breed and litters are small.

CHOICE ONE

Pass the proposed law:

a. Increase the bounty or reward from $2 to $5 for killing a coyote.
b. Try to kill all coyotes in livestock- and poultry-raising areas. Specialists in extermination would come from the Fish and Wildlife Service of the federal government. They would put an extremely powerful poison in meat and leave it where coyotes would find it.
c. Open all government lands to year-round hunting of coyotes.

CHOICE TWO

Defeat or do not vote on the proposed law. This leaves the bounty at $2 and allows farmers and ranchers to kill coyotes if they want to.

CHOICE THREE

Make changes in the proposed law and pass the new one.

You may omit any of the present provisions and add one or more new ones from the following list:

a. Raise the bounty to $25.
b. Have the Fish and Wildlife Service kill all coyotes in the state with poison.
c. Employ state wildlife specialists to help farmers get rid of killer coyotes.
d. Permit hunting of coyotes only when the coyote population is high, and rabbit and rodent populations are low.

Mastery Item 14-1

Managing a lake

Mr. and Mrs. Paul Jackson just bought Round Lake. The lake does not contain any large fish that are exciting to catch. But there are some medium-sized fish, such as bluegills and crappies.

FIGURE 14.9
This drawing shows the food web in Round Lake now.

Producers

First-order consumers

Second-order consumers

FIGURE 14.10
The food web in Round Lake would look like this if the Jacksons added new fish to the lake.

Around the lake shore there are cottages and boats to rent and a store that sells bait and food to the vacationers.

The Jacksons want to attract more fishermen to the lake. They have decided to add new species of fish to the lake. These new fish grow to a large size and are exciting to catch. The food web in Round Lake now is shown in Figure 14.9, on the preceding page.

After the new fish are added to the lake, the food web will be like Figure 14.10.

The Jacksons assume that the lake will support as many medium-sized fish as before, plus a good number of the new large fish. They think more fishermen will want to fish in Round Lake. They think that families who already came to fish for the medium-sized fish will still be satisfied, and will continue to come fish in the lake.

What do you think will happen with the new fish in the lake? Write a statement to the Jacksons. Tell them whether their plan is likely to work. Give reasons to support your position.

Key

You should think twice before trying to change food webs. The Jacksons should be warned. Many medium-sized fish may be eaten by the large fish. As a result, the family fishing would probably decrease. If too many big fish are added, they might eat most of the little fish. Less food would mean fewer large fish, thus fewer fishermen. Then everyone would be dissatisfied.

Mastery Item 14-2

A case history in managing wildlife

Here is a problem that involves predators and their prey. Can you suggest a good solution to the problem?

Part of the **Kaibab Plateau** (KYE-bab pla-TOH) in Arizona is shown in Figure 14.11. The plateau is covered mostly

FIGURE 14.11
There are steep drops around the Kaibab Plateau. Deer and other animals cannot freely migrate to other feeding grounds.

by evergreen forests and grasslands. The plateau is surrounded by rugged canyons, including the Grand Canyon. They form natural barriers to the movement of most animals. The animals on the plateau include deer, mountain lions, coyotes, bobcats, rabbits, ground squirrels, chipmunks, and wood rats. (See Figure 14.12.)

Mountain lions are mainly deer eaters. Coyotes and bobcats eat some deer, but depend on rabbits and the small rodents for most of their food. The rodents and rabbits eat grasses and low shrubs. The deer eat these, too. They also eat the buds and tender branches of trees in their reach.

Before the arrival of the white man in the late 1800's, Indians lived on the plateau. They killed about 800 deer each year for food and deer hides. The Indians seldom killed any of the predators (mountain lions, coyotes, or bobcats). During this time the deer population probably did not change greatly from year to year.

A study of the plateau animals was made in 1906. The deer population was estimated to be about 4,000. About this

FIGURE 14.12
A food web on the Kaibab Plateau

time, President Theodore Roosevelt made the plateau into a large game preserve. The killing of deer was prohibited. A campaign was begun to kill the predators. As a result, about 10,000 predators were killed between 1906 and 1939.

Estimates of the deer population were made during these years. By 1924 there were about 100,000 deer! However, much of the plant life had been eaten or destroyed by then. Thousands of deer died of starvation or disease during the next few years. The number of deer was down to about 30,000 by 1930.

Suppose you were called in by government officials. They need advice on the management of Kaibab Plateau's wildlife. How many deer would you suggest the plateau could reasonably support? (This number should not destroy much of the plant life, nor should many deer starve.) What actions would you suggest to keep the deer population at the size you suggested? Give reasons for your answers to both of these questions.

Key

The plateau area can probably support at least 4,000 deer. This number had lived there for many years without overeating the plants. To maintain the deer population at about 4,000, fewer predators should be killed. Also, some hunting could be permitted to remove excess deer. The surrounding canyons make it difficult for excess deer to migrate or be transported to other areas away from the plateau. Another solution might be not to kill *any* predators. You could allow hunting if either predator or prey increase in numbers over a period of a few years.

FIGURE 15.1
What water looks like under a microscope, as drawn by a nineteenth century cartoonist

Investigation

15 Competing with microbes

Organisms that live in the same environment often compete for food and living space. For example, men and coyotes compete in Kansas. In this Investigation you will gather data on man's most common competitors: the microbes.

Microbes (MY-krohbs) are small organisms, so small that the individuals can be seen only with a microscope. You looked at some aquatic microbes in previous Investigations. How could such small organisms compete with man? In Problems 15-1 and 15-2, you will investigate that question.

Then in Problem 15-3 you will investigate another aspect of microbes: the way they prevent the world from turning into a giant, stinking garbage dump.

Problem 15-1 — Why does food rot?

You know that food will spoil or rot if it is improperly cared for or stored too long. Doing this Problem should help you understand why food rots.

Materials
fresh hamburger
2 test tubes with plugs

test tube holder and rack
pyrex beaker or metal pan
Bunsen burner
glass-marking pencils
microscope
microscope slide
coverslip
medicine dropper or wire loop

Gathering data

Get two test tubes and two pieces of hamburger each about one-half inch square. Put the same amount of hamburger in each tube. Shape the pieces so that they slide easily to the bottom of each tube. Then fill each tube half full of water. Plug the tubes and mark them so you can identify them later.

Place one tube in boiling water for 15 minutes as shown in Figure 15.2. Then remove the hot tube with a test tube

FIGURE 15.2
Place one test tube with hamburger in a water bath for 15 minutes. Do you know what this boiling will do?

holder. Place both tubes in the same test tube rack in a warm place. Mark the boiled tube number 1, and the other tube number 2.

Observe the appearance of the meat and the liquid in each tube every day for three days. Look for such things as changes in color, texture, or odor. On the third day remove the plugs to make your final observations.

Use the technique you learned in Investigation 9 to make a separate wet mount of a drop of water from each tube. Observe each drop carefully under the high power of the microscope. If you reduce the amount of light, it may help your observations. Do both drops look the same?

Recording data

Record your observations of each test tube each day. Draw or describe in writing any differences you can see under the microscope on the last day.

Analyzing data

What inference can you make from your data? Does the data help answer these questions: Why does food rot? How could you apply what you have learned to food handling and preparation at home? In restaurants?

Problem 15-2

Do microbes cost you money?

Almost all of the "fresh foods" we eat are picked, dug up, or butchered *days* before we buy them. Only 5.4 per cent of Americans live on a farm. Everyone else gets his food shipped in, which takes time. Of course, farmers go to the store for some of their food, too.

A good part of the food our farmers and ranchers produce is not grown to be sold fresh. For example, some food grown in the summer must be stored for winter eating. Some food is grown to be a surplus supply. The government buys it from farmers. Part of this surplus food is sold or given to people who cannot afford to buy it. Part is sold or given to other countries to relieve starvation.

FIGURE 15.3
Ask the grocery manager how his store keeps foods from spoiling What does he do to make sure the food his customers buy is safe to eat?

Why don't microbes spoil all our food as quickly as they spoiled the hamburger in Problem 15-1? You are going to gather data to help answer this question. It is a tremendous battle to keep food from rotting before man finally gets to eat it!

Gathering data

The people who manage the grocery store in your neighborhood are experts on food spoilage. You will gather data from them on preserving food. See Figure 15.3. No more than four students should gather data from each store. Pick a time when the store is not very busy. Interview the manager. Tell him you are doing a project for school on how to preserve food. If the store has a "meat man," a baker, and a "vegetable man," talk to them, too. Each interview should take about 10 minutes.

Ask the following kinds of questions (or make up your own).

GENERAL QUESTIONS FOR THE MANAGER

a. Which foods can you keep the longest? Which foods have the shortest storage times?
b. What ways are there to keep foods from spoiling?
c. Is there any law that tells you how long you can keep food in the store and still sell it?
d. Do you ever have to send some foods back to the supplier? Why?
e. Do other kinds of food besides vegetables become unusable if they are in the store too long?
f. Do you have any idea how much it costs the store each week to keep food from spoiling? Or how much food has to be thrown out?
g. When food spoils and is thrown out, how is this loss paid for?

QUESTIONS FOR MEAT, VEGETABLE, AND BAKERY SPECIALISTS

a. How long can you keep meat, vegetables, fruit, or baked goods?
b. What kinds of items do you have to sell most rapidly?
c. What special ways do you have to keep foods from spoiling?
d. Why is it difficult to keep meat, vegetables, fruit, or baked goods?

Analyzing data

You should be able to discuss these topics in class. Use the data gathered in your survey.

a. Why can some foods be stored longer than others?
b. What are several ways to prepare foods so they stay edible?
c. Does the cost of operating a store include losses from spoiled food? Who pays for this loss?
d. How does the local grocery store insure that its customers get unspoiled food?

Problem 15-3

What happens to garbage and trash?

Many communities in the United States collect waste materials from their citizens and then dump it in holes in the ground. Everything is dumped, from kitchen garbage, to old toys, to junked cars. Then a bulldozer covers it with dirt. What happens to all these materials? As the world population grows larger, we produce more and more waste materials. Can we keep on using these dumps? In Problem 15-3, you will gather data to help answer these questions.

Materials
junk and garbage

Gathering data
Bring to school a piece of junk or garbage about one inch square. Your group should test as many materials as possible. Include discarded food, several kinds of metal, glass, and plastic. Place the materials on a layer of wet dirt, as shown in Figure 15.4. Carefully record the color, size, and texture of each kind of material. Sketches or photographs of your experiment would be best.

Cover the materials with an inch or so of damp soil and set the container in a warm place. Let it set for a week. Don't let the soil dry out.

After one week, carefully remove the top layer of soil and record the condition of the materials. If you want to do this experiment for a longer time, ask your teacher.

Analyzing data
Which of your materials changed? What do you think caused the changes? Which of the materials did not seem to change?

If the materials you tested are found in city trash, do you think that cities will be able to use dumps over and over?

Can you make any suggestions to help prevent the world from being covered with junk? What can your family do to keep down the amount of junk to be buried?

FIGURE 15.4
Test several different kinds of garbage by burying them in damp soil. Which materials changed after a week?

Mastery Item 15-1

A bulging can on the shelf

Suppose you are helping a friend shop for groceries. He picks a can of vegetables from the shelf. The can is bulging and swelled up at the ends. What advice would you give him about buying it? What reasons would you give for your advice?

Key

You should advise him not to buy the can of vegetables. The bulging ends suggest that the food inside has not been properly sterilized. Microbes are using the vegetables for food.

Mastery Item 15-2

Eating a woolly mammoth

In 1806 a scientist in Siberia dug the body of a woolly mammoth out of frozen ground. A woolly mammoth is shown in Figure 15.6. Steaks from this 50,000 year old animal were eaten by men and dogs.

Make a hypothesis to explain why this mammoth was still edible after being dead for 50,000 years.

Key

The woolly mammoth was preserved at such a low temperature that the growth of microbes was prevented, so the meat didn't spoil.

FIGURE 15.5
Workers discovered the remains of a 250,000 year old woolly mammoth while they were doing roadwork in Rome.

FIGURE 15.6
This picture of a woolly mammoth appears on a mural in the American Museum of Natural History.

Mastery Item 15-3

Do we need a "super-preservative"?

Suppose you were a chemist who discovered a new chemical. Any living thing sprayed with this chemical would not rot when it died. Even animals that ate treated plants and animals would not rot. The chemical stops microbes from growing. The chemical is harmless in every other way.

The leaders of several large countries want to know if they should add this chemical to foods. It could increase their food supply by 20 per cent. They are even thinking of spraying all the farmlands in their countries with it. What advice would you give the leaders? Would you sell them your chemical?

Key

The chemical could help increase the world food supply. If plants and animals did not rot, they could also easily be shipped to hungry peoples. The chemical also could cause the surface of the earth to become cluttered with unwanted, dead organisms and other materials. There would be no microbes to break them down. Until you found out how to reverse its effects, the chemical should not be widely used.

FIGURE 16.1
Is the Earth becoming too crowded? What will happen as population continues to increase?

Investigation

16 Planet management

How many people can live on Earth? Is there a limit? No one knows for sure.

Almost 200 years ago Thomas Malthus predicted that a limit would be reached *soon*. Population would increase until food ran out. Then many people would starve to death. Malthus didn't know that food production would be greatly increased by improving farming. Some people say his prediction will still come true—only later than he said.

Some other people say that food production isn't quite that important. They point to India. It has a general food shortage and *also* has one of the fastest growing populations on the Earth.

The growth of human populations is not well understood. One reason is that human populations can't be experimented with in a laboratory. However, in this Investigation you will experiment with a population of beings on the planet Clarion. Actually, the experiments you do will be *realistic* (similar to reality), but not real. Clarion is an imaginary planet.

Your experiments will resemble conditions that might occur on Earth. As you do these experiments, you will learn about populations. In some ways, the experiments will be like a game. You will probably find the experiments, or game, more interesting if you use your imagination. The more realistic Clarion can be for you, the more you will learn and the more you will enjoy the experiments.

Problem 16-1

How can Clarion be improved?

Imagine that you are living in the year 2000 A.D. You have a high-paying job with Universal Planet Management Associates. You and your team have been given the task of managing the planet Clarion for the next 50 years. (A Clarion year is much shorter than an Earth year.)

As "Manager for Clarion," you have to decide how to spend Clarion's improvement budget. You will set up a new improvement budget every five Clarion years. At the end of each five year period, you will find out how the planet has changed during that time. You will find out about population change, incomes, the food supply, and the condition of the environment. You can use this information to decide on your next five year budget.

What Clarion is like

The planet of Clarion is very much like the Earth in size, climate, and geography. However, the land area is smaller than the Earth's. The living things on the planet are similar to plants and animals on Earth with one exception—there is no native animal like man.

About 10 years ago the Council of Civilized Planets transported 100,000 colonists to Clarion from the Planet Rasmuss. These colonists are similar to people, but are smaller. They have shorter life spans than people and look a little different, too. You could call them "humanoids." Their foods are similar to ours.

The natural state of Clarion is like Earth's before man began to change it. Some of the land is flat and has a mild climate that is good for farming. There are some deserts of purple sand. They are fiery hot during the day and cold at night. There are other regions of bare rock that separate the fertile areas. All the land is on one continent. The rest of the planet is a blue-green ocean much less salty than Earth's.

There seems to be only one major threat to life on Clarion—the disease Holobinkitis. It attacks both plants and animals, Clarionmen included. No one knows the cause of

FIGURE 16.2
Before the humanoids arrived, the planet Clarion looked very much like Earth.

the disease or a cure for it. It is almost always fatal if you catch it.

Clarion has one large and unusual mineral deposit. The mineral is Walterite ore. In fact, Clarion is the only known planet with such large deposits. Several scientists say that the metal obtained from the ore is superior to iron or steel for most uses. However, Walterite is usually found under a thick layer of hard rock. Thus, it is difficult to mine.

In comparison to Earth, Clarion has a low standard of living. The Clarionmen don't yet produce all the things they need and want. As a result, they are thrifty. They do not like to spend money unless it's absolutely necessary. Then, they want to get a "good deal" for every bit of money they spend.

They have some of the same problems the Pilgrims had when they first settled in North America. Food sometimes becomes scarce, and it is expensive to import food from other planets. One of the big problems with raising crops is that the insects on Clarion really like to eat them. Therefore, a large part of every crop is lost to insects.

FIGURE 16.3
The Clarionmen made many changes on the planet. Do you think the planet benefited from these changes?

Most of the Clarionmen say they will stay on in spite of the hardships. They believe Clarion has a bright future once things get going. The Clarionmen usually give three reasons for leaving their home planet of Rasmuss:

1. It was getting too crowded there.
2. There was a promise of a better life on Clarion.
3. Clarion is a beautiful place to live.

Most Clarionmen have a strong appreciation for beauty, and they enjoy Nature. They look forward to more leisure time outdoors.

Materials

The Planet Management Game™, including:
Planetary Status Ledger
Project List and Cards

4 colored marking pencils
4 graphs
penny

Playing The Planet Management Game™

How will you manage affairs on this planet? Can you and your team do a better job than others playing the game?

The whole game lasts 10 rounds. Each round represents five years on Clarion. In each round you spend money on certain projects. The object of the game is to get Clarion to the most desirable status at the end of 50 years.

THE PLANETARY STATUS LEDGER

Before starting, you should understand the Planetary Status Ledger. It is something like a scorecard for the game. One is shown on the next two pages. There is a place to record how you spend your money, and a place to record changes on Clarion. Change is measured on these four indexes.

The **Population index** begins at 100. Each unit stands for 1,000 Clarionmen as the game begins. Imagine that after the first round, the population index changes to 104. This means the population has actually increased by 4,000.

The **Food index** begins at 100 units. Each unit stands for enough food for a well-balanced, but simple diet for 1,000 Clarionmen. (Everyone has just enough to eat at first.) In the future there may be less than one unit of food per 1,000 citizens. Then, some people will be under-fed or maybe even starving. If there were more food units than Clarionmen, the meals could be fancier.

The **Income index** begins at 100 bux. This column shows all the money earned by all the Clarionmen in one year. Therefore, it also measures how much they have to spend for food, clothing, recreation, and so on. Imagine that the income index goes up to 120, while the population index stays at 100. This means the average income of individuals has risen. It probably means a higher standard of living for Clarion. However, income might stay at 100, and the population index go up to 120. Then the situation would be reversed. More Clarionmen would be sharing the same amount of income.

The **Environment index** measures the quality of Clarion's environment. The index was at 100 when the Clarionmen arrived. Then the air was pure. The rivers and lakes were clean. The land was very beautiful. However, the environment index at the beginning of the game is 90. The difference of 10 points has occurred because of the effects of colonizing the planet Clarion.

FIGURE 16.4 *This is a sample table. Don't write on it.*

PLANETARY STATUS LEDGER—PROJECT RECORD

Round	Amount spent					
Begins in 2000 A.D.	red	blue	orange	yellow	green	Total
1						
2						
3						
4						
5						
6						
7						
8						
9						
10 ends at 2050 A.D.						

Total money spent _____

PLANETARY STATUS LEDGER—INDEXES

Round		Population	Income	Food	Environment
begins in 2000 A.D.		100	100	100	90
1	change				
	new index				
2	change				
	new index				
3	change				
	new index				
4	change				
	new index				
5	change				
	new index				
6	change				
	new index				
7	change				
	new index				
8	change				
	new index				
9	change				
	new index				
10 ends at 2050 A.D.	change				
	new index				

The rules

After you play one or two rounds, you will find these rules quick and easy to follow.

1. Read the Project Guide and select the projects you'd like for Clarion. (You can pick only one project from a color group.) You may spend 2, 4, or 6 bux on any project you choose. One buc is worth about one million dollars.

    You can't spend more than 10 bux total in any round. However, you may choose to spend less.
    Example: You choose:
    Project B—2 bux
    Project H—2 bux
    Project K—6 bux
    None in yellow color group
    None in green color group
2. Select the cards that match the projects and the *amounts* spent on each.
3. Record your selections and the amount spent on each in the Planetary Status Ledger. Here is an example:

PLANETARY STATUS LEDGER—PROJECT RECORD

Round	Amount spent					
	red	blue	orange	yellow	green	Total
1	B-2	H-2	K-6	M-0	R-0	10

4. Place the Project Cards together. One hole (and only one hole) should open through all of them.
5. Turn to the data sheets labeled Rounds 1 and 2, POPULATION CHANGE. There are four sheets in the set, numbered 1–4.

    Flip a penny twice to decide which data sheet to use. Use this guide:

First flip	Second flip	Use data sheet
heads	heads	1
heads	tails	2
tails	heads	3
tails	tails	4

6. Place the group of Project Cards on the data sheet and read the results through the open hole.
   Example: You flipped heads first and tails second. This gives data sheet 2.
   Placing the cards on data sheet 2, a "4" shows through. Record the change in population, 4, on the Planetary Status Ledger.
7. Add the change to the population index and write the results on the new index line. If the change was a minus number, it would be subtracted.
   Example: Starting with 100 and adding 4, gives 104 as the new index.
8. Repeat steps 5 and 6 using INCOME CHANGE. Then do the same with FOOD CHANGE and ENVIRONMENT CHANGE. Use only the data sheets labeled Rounds 1 and 2. When you have recorded the changes for all four indexes, you have finished Round 1. Place the Project Cards back in the pack.
9. To play Round 2, pick a set of projects and proceed as you did in Round 1.
10. As you play Rounds 3-10, be sure to turn to the right data sheets. You might want to keep graphs of changes on each index. That might help you to see the results of your projects.

The game is completed at the end of 10 rounds.

Do you think the Clarionmen would be pleased with the state of their planet at the end?

Did you manage Clarion well?

Could you have accomplished the same results with less money?

Discuss with other Planet Management teams what the ideal conditions on Clarion should be after 50 years of management. Once this is agreed upon, the winner of the game will be the person or team who came closest to the ideal conditions. Did you win?

You might want to play "The Planet Management Game^{TM}" again, now that everyone has agreed on the ideal conditions for the planet Clarion.

Will you do the same projects?
Will you be able to save money?
How good can you become as a Planet Manager?

FIGURE 16.5
Which projects would you spend your bux on as Manager of Clarion?

Project Guide

RED CARDS

A = No project in this color group.
B = Clear more land and prepare it for raising crops.
C = Import chemical fertilizers and pesticides.
D = Buy machinery to plant, cultivate, and harvest crops.

BLUE CARDS

E = No project in this color group.
F = Conduct research to develop new varieties of corn, wheat, and rice with higher yields per acre.
G = Conduct research to find a cure for Holobinkitis.
H = Conduct research to find better ways to mine and use Walterite.

ORANGE CARDS

I = No project in this color group
J = Build more Walterite mines and purification plants.
K = Build more highways and railroads to connect towns.
L = Construct new electric power plants.

YELLOW CARDS

M = No project in this color group.
N = Build more schools, for all age groups and for general, professional, and technical education.
P = Pay people to go to graduate schools of civil engineering, biochemistry, medicine, and agriculture.
Q = Finance the "Einstein Plan." This would a. send a few of the most capable individuals to Earth for advanced study, and b. connect more Clarion homes to the educational television network.

GREEN CARDS

R = No project in this color group.
S = Construct more and improved sewage treatment plants for city and factory wastes.
T = Finance the Wright Plan: a. buy areas of land around lakes and oceans to be used for recreation, b. build mile-high apartment buildings surrounded by 100-acre parks in the center of cities.
U = Equip more chimneys of homes, industries, and other buildings with antismoke equipment.

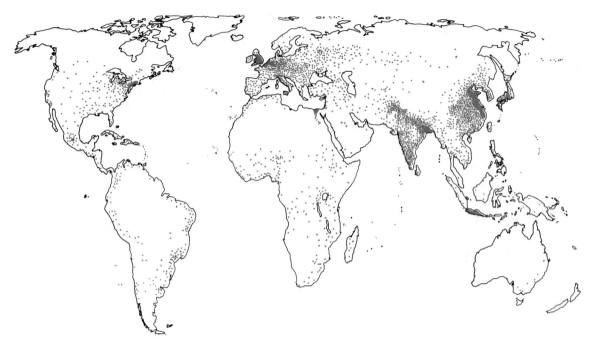

FIGURE 16.6
This map shows which areas in the world have the most population. Each dot represents 750,000 people. Areas that are solid green have very large populations. Does this map suggest any indexes to use for an Earth Management Game?

Mastery Item 16-1

The Earth Management Game

If you were making an Earth Management Game, what indexes would you plan on using? What are your reasons for the indexes you listed?

Key

Your answer is acceptable if you listed:
a. four or more indexes.
b. one of which was not used in Planet Management, and
c. you had a definite reason for your answer.

Planet management

Mastery Item 16-2

Improving The Planet Management Game™

Suggest how to improve the Project Guide so players can reach better final results.

Key

At least one of your suggestions should be a project to directly limit the number of humanoids on Clarion.

Mastery Item 16-3

Planet management and Earth today

What two situations on Earth today are like situations on Clarion? Give reasons for your answers.

Key

Many answers are possible. If your answers aren't included below, check with your teacher. Your answer might deal with:

a. The shortage of food in some parts of the world.
b. Unused natural resources in various parts of the world, for example iron ore and petroleum.
c. Uncontrolled and serious diseases in some parts of the world, for example malaria.
d. The pollution of the environment, especially in crowded areas with much industry.
e. The crowded conditions in some parts of the world.

FIGURE 17.1
What use is the skunk's spray?

Investigation

17 Organism versus organism

Observe the dishes of seeds your teacher shows you. These seeds were planted on the date recorded on the containers. Discuss your observations with your classmates. What inferences or hypotheses can you make to explain what you see?

Problem 17-1

Offense and defense

Think about what happened in the dish with two kinds of seeds. Do you think you would get the same results if you mixed other kinds of seeds? What would happen if other kinds of living things were placed together? Do you think they would interact in a similar way? In this Problem your class will try to answer these questions.

The Materials list suggests organisms you could investigate. Your teacher will show you several examples of these groups of organisms. You may think of other organisms to bring in that you would rather use.

Materials

seeds   bacteria
roots   people
mold

Gathering data

The organisms in the Materials list are described below. These descriptions will help you decide which living things to use in your investigation. Choose a combination of organisms to study. Try to pick the kinds of organisms that generally live in the same environment.

Some suggested procedures to obtain data are also described below. *You don't have to read these procedures right now.* Wait until you pick the organisms you are interested in. Then read the procedures. You can use them or think up your own.

Design an experiment to help you test your hypotheses about organisms affecting one another. Discuss your plan of action with your teacher before you start.

SEEDS

Most seeds in nature are found in or on the soil. You have already discussed the results of planting two kinds of seeds together. Will other seeds show the same results as the seeds you saw before? What if you plant them in sand or soil instead of on paper towels? What other organisms might interact with a seed in the ground? Another question is, can you find out what part of a seed might affect other seeds?

Suggested Procedures. All seeds can be tested the same way as the seeds your teacher showed you. You can also test substances from inside seeds, if the substances dissolve in water. To remove these substances, soak about 30 gm of seeds in 150 ml of tap water. Small seeds, like Merion bluegrass, should be soaked for about 12 hours, large seeds for 24 hours. Stir them occasionally during the soaking period. Afterwards, filter the liquid through a clean piece of nylon cut from an old stocking.

You can test several kinds of seeds at the same time in different areas of a Petri dish. Add to the dish the water in which you soaked your seeds.

ROOTS

Do live roots interact with other organisms? Do they affect the germination of seeds? Will molds or bacteria grow around and on a slice of root? Figure 17.2 shows some of the many kinds of organisms found in the soil. How do roots interact with soil organisms?

FIGURE 17.2
Scientists have found that forest soil teems with life. One square foot of soil one inch deep can house thousands of worms and insects and billions of bacteria, molds, algae, and other microbes.

Suggested Procedures. You can use pieces of roots you dig up. Or you can cut up roots your teacher provides. Wash the roots well before using them.

To see how seeds react to roots, place the seeds around slices of roots in a Petri dish. You can also test the effects of root juice. Crush about 30 g of roots in a container with 100 ml of tap water. Shake the mixture well. Then strain the roots out of the juice through a *clean* piece of old nylon stocking. Add the juice, instead of water, to the dish with seeds you are testing. You can observe several kinds of seeds at the same time in different areas of the dish.

You can test the effect of roots on bacteria and molds. You can grow the microbes on **agar** (AH-gar), a special jellylike food. First, swab dishes of agar with the microbes. (See procedures in the section on bacteria.) Then, gently place slices of roots on the agar's surface.

MOLDS

Molds can't make their own food. You learned in Competing with Microbes that they often eat dead organisms. They are usually found in dark, moist places in the soil. You may have seen them growing on old bread or fruit. When you investigated seed germination, molds may have grown on your seeds. You can also grow them in the lab on agar.

Do molds and bacteria affect each other? Are molds affected by seeds, roots, and even humans?

Suggested Procedures. Ways to study interactions between molds and other organisms are described in the next section on bacteria.

BACTERIA

Bacteria are microbes present nearly everywhere. They are in the air, soil, water, on and in all organisms. The bacteria used in this Investigation are usually found in the soil. Sometimes they are found in the food you eat. You saw other bacteria rot food and garbage in Investigation 15.

Bacteria, like molds, grow well on agar. If you spread a tiny amount of bacteria over a dish of agar, soon you can see colonies grow. Each colony contains millions of bacteria, which may grow over the entire surface.

Remember that some bacteria and molds cause disease or infection. None of the kinds used in the class cause infection. But when working with microbes, you should not take chances. *Everyone in the class must follow these rules.* They will help prevent unwanted microbes from getting into experiments.

1. Do not eat or drink anything in the classroom.
2. Never put things in your mouth while you work in the lab.
3. If you spill a culture of bacteria or mold, call your teacher to help you clean up.
4. Always wash your hands after class.

Suggested Procedures. To grow and test bacteria and molds in Petri dishes you will need:

sterile Petri dish   sterile cotton swabs
test tube of sterile agar   Bunsen burner or alcohol lamp
at least 2 kinds of microbes already growing in tubes of
      liquid

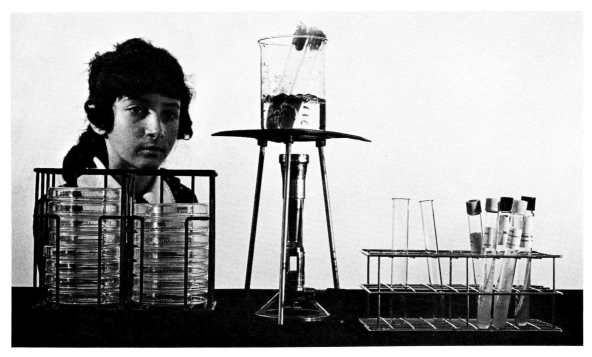

FIGURE 17.3
Follow the procedure described in the text to prepare agar for growing bacteria and molds.

First, melt the tube of sterile agar in a water bath, as shown in Figure 17.3 *top*. Remove the plug. Then flame the open mouth of the tube for two or three seconds to sterilize it, as in Figure 17.3 *bottom*. Pour the agar into the dish as

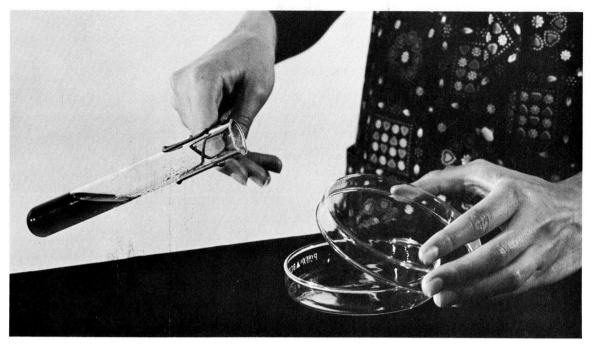

FIGURE 17.4
The sterile Petri dish should be recovered as soon as the agar is poured into it. Can you think of a reason why?

shown in Figure 17.4. Tilt the dish just enough to make the agar cover the bottom. The agar will harden in about five minutes.

Next, take a tube of microbes. Tap the tube with your fingers to scatter the organisms which may have settled to the bottom. Avoid getting the plug wet. Ask someone to remove and hold the plug for you. You can do the rest.

Sterilize the mouth of the tube. Dip a cotton swab into the liquid. Then reheat the top of the tube and replug it. Now gently swab the liquid over the whole surface of the agar as shown in Figure 17.5. Be careful not to break through the surface. Immediately discard the swab in the container labeled "Used Swabs."

Using the same technique as before, unplug and sterilize the mouth of a tube that contains *another* kind of organism. Wet another sterile swab and replug the tube. Gently touch the center of the agar with the swab. Make certain some of the liquid comes off. Discard the swab in the "Used Swab" container. What will you use for a control?

If you want to test the actions of other organisms on each other, ask your teacher for more materials.

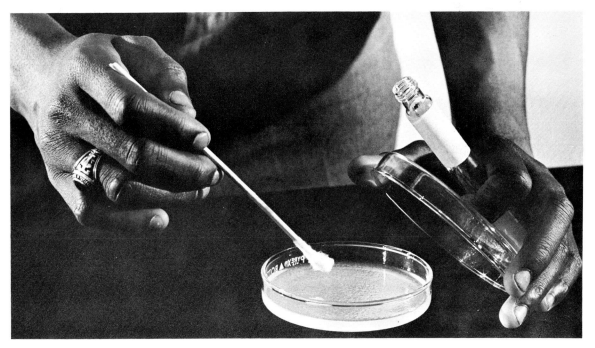

FIGURE 17.5
Transfer the microbes on the wet swab to the agar.

HUMANS

You can choose to be one of the test organisms. Is your body the kind of environment that some microbes need? Your eyes and mouth are places where microbes can enter your body. Some can also live there. How does your body defend itself against microbes? Do you think tears or saliva might affect microbes?

Suggested Procedures. You can test the effect of tears or saliva on bacteria and molds. For example, you can mix tears with bacteria and observe the results. If you choose this method, here is what you will need:

piece of fresh onion
several small sterile
 test tubes
several kinds of bacteria
medicine droppers
an incubator or heater that will keep a temperature of 110°–115°F (43°–46°C).

Collect tears from a classmate. Have him hold a slice of onion near his nose. Then you can catch the tears in the small test tubes.

FIGURE 17.6
This student is using an onion to make her classmate cry. What will she do with the tears she is about to collect?

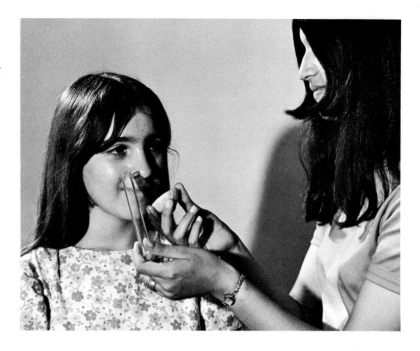

Collect seven or eight tears in a test tube for each kind of bacteria you will test. Label each test tube with the name of the kind of bacteria you plan to use. Then add 1 ml (20 drops) of bacteria to each tube. Plug the tubes. Gently shake them to mix the bacteria and tears together. What will you put in the control tubes?

Then mix hot and cold water in a small jar to get a temperature between 110°–115°F. Put your tubes in the jar of water. Place the jar in an incubator at 110°–115°F (43°–46°C) for 15–20 minutes.

Now observe the liquid in each tube. Does it look less cloudy? Is it thicker? If you don't see any change, leave the tubes in the incubator until tomorrow. If possible, check them several times before then.

If you want to work with saliva, use procedures similar to those described for tears. Rinse your mouth with water before collecting the saliva. What is the reason for rinsing your mouth?

Recording data

Write down your procedures and your observations. You might want to describe your observations in a series of drawings. If you grow mold or bacteria, your drawings should show all growth in the dishes. If you have a camera, you can take pictures to show the changes. A camera that can take pictures from as close as one foot is necessary to show small, but important details.

Analyzing data

Discuss the results of your own investigation and the ones your classmates did. Review the hypotheses and questions that led to your investigation. Did you find any answers?

What kinds of interactions were produced? How might the interactions that you observed benefit the organisms? Does their behavior help them survive? Many of the kinds of interactions you and your classmates have observed are called **antibiosis** (an-tee-by-OH-sis). Can you think of any other examples of antibiosis?

Mastery Item 17-1

When the tide turns red

The sea is not always blue. Along the Florida coast of the Gulf of Mexico, the sea sometimes becomes red. To the people who live along this coast this "red tide" has a special meaning. When the red color appears, they can expect thousands of fish to die soon and wash onto the beaches.

Biologists who examine the water discover it is alive with microbes with red eyespots. These eyespots give the sea its reddish color.

The biologists nicknamed the organism "Jim Brevis," after its scientific name, **Gymnodinium brevis** (Jim-no-DIN-ee-um BREV-is).

Was Jim Brevis responsible for the fish killed? If so, were the deaths caused by antibiosis? Suppose you could conduct experiments to test the effects of Jim Brevis on healthy fish. Describe an experiment you would design to discover whether fish are killed by a substance from Jim Brevis. What data would show that such a substance was the cause of death?

FIGURE 17.7
What data would you gather to find out if the fish kill was caused by the "red tide"?

Key

One experiment would be to observe healthy fish in two containers of ocean water. One container would have water from which Jim Brevis had been removed. The water in the other container would not have contained Jim Brevis. Otherwise the water would be the same. If the fish only in the Jim Brevis water died, this would indicate that antibiosis was the cause of death. You may have designed other experiments and described other evidence. Discuss your answer with your teacher.

Mastery Item 17-2

Turnips and marigolds

A gardener in Maine had a problem. For several years something in the soil had ruined most of his turnips. Nearby he had a bed of healthy-looking marigolds. These flowers have a strong odor. He checked their roots but found almost no signs of damage.

The gardener took samples of soil from the turnip patch and marigold bed. He sent them to the Agriculture Department for analysis. Both samples were similar except for the number of **nematodes** (NEM-uh-toads). These are tiny worms that may live in the soil and harm roots. The turnip soil had 10 times more nematodes than the marigold soil. The nematodes had ruined the turnip crops.

Write a hypothesis that might explain why there were fewer nematodes in the marigold soil. Also suggest a way to save the gardener's next turnip crop. Explain why you think it will work.

Key

The marigold roots could have given off a substance that killed nematodes or kept them away.

The gardener could move his turnip patch to the marigold bed or try planting the two together. Any substance from marigold roots that harmed nematodes would keep them away from the turnips.

You may have a different hypothesis and plan, so check them with your teacher.

FIGURE 18.1
The redwood park issue was debated in the newspapers and the Senate Chamber.

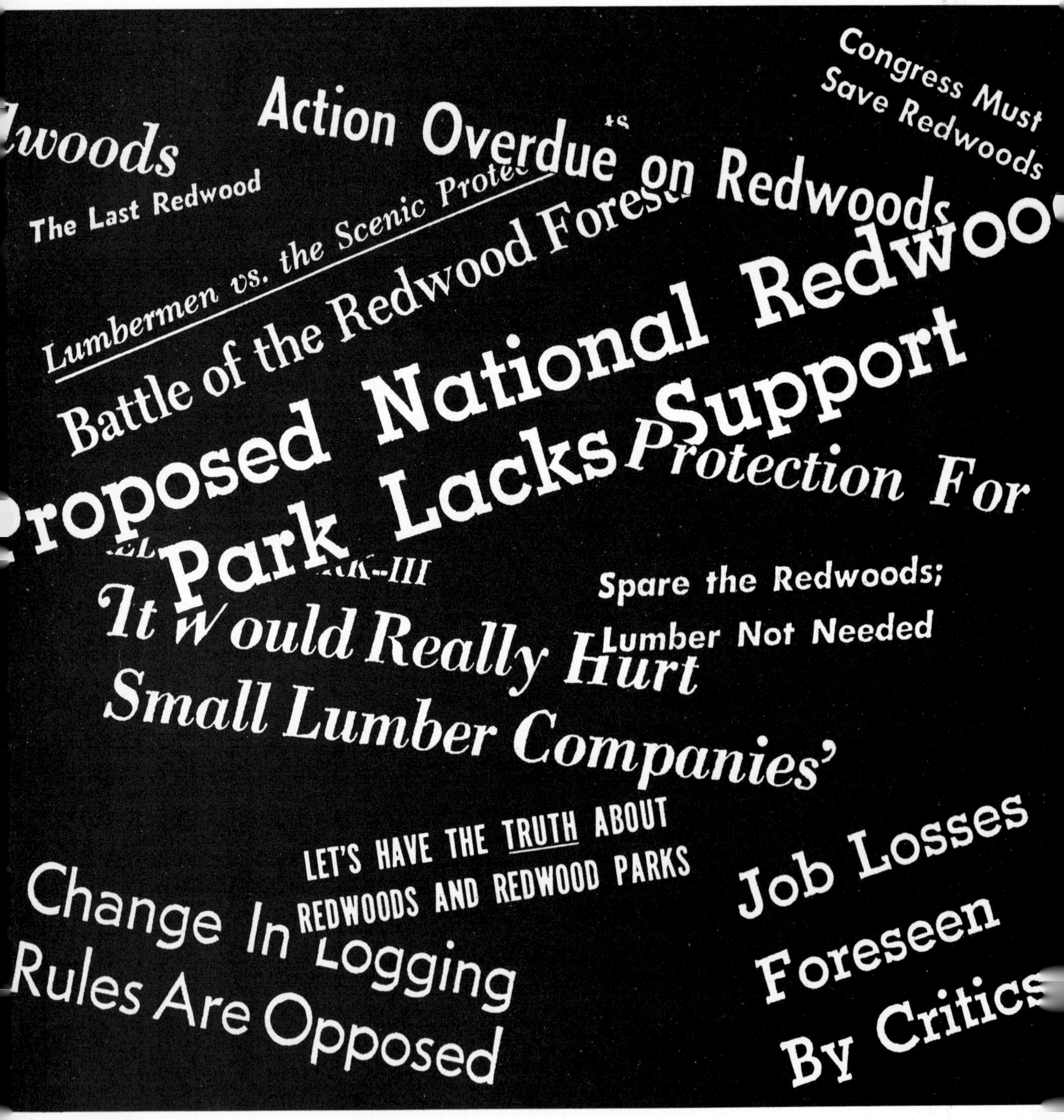

Investigation

18 The Redwood Controversy™

Many parts of the country are already crowded. But by the year 2000 the population of the United States may double. Imagine what it will be like to have twice as many people living in your area!

If there were world enough, everyone could own as much land as he wanted to. But there isn't. If we had an endless supply of natural treasures—plants, animals, and unspoiled countrysides—we could afford to use up some of the supply. But we can't.

How do we decide who shall own the land and how the land shall be used?

In this Investigation you will play a game called "The Redwood Controversy™." You will learn about one way land use problems are solved. You and your classmates will act out the parts of United States Senators and leading citizens. The classroom will be turned into the Senate Chamber.

Problem 18-1

What will be the fate of the redwoods?

Your class will conduct a Senate hearing to decide whether to establish a Redwood National Park. The roles you will play are not real people, but the evidence you will listen to is real. The kind of hearing you will recreate actually happened. Only this time *you* have to decide what to do.

Materials
"The Redwood Controversy™" a game.

The plot of the play
The United States Senate is about to consider setting up a Redwood National Park. Conservationists have tried to put such a bill in front of the Senate for almost 70 years.

All over the United States, people are watching this Senate meeting. Many newspapers support the proposed park. Lumber companies and people who work in the redwood country strongly oppose any park. Whatever decision is made, there will be many unhappy people. It is the responsibility of each Senator to carefully consider all of the data and opinions. He should listen to all of the witnesses. Then he must vote for one of the following proposals.

Proposal #1

Establish a **small national park** by taking over two state parks plus 5,760 acres of private land. Estimated cost to the U.S. government to purchase the private land is 6 million dollars.

Proposal #2

Establish a **medium-size national park** by taking over three state parks plus 18,900 acres of private land. Estimated cost to the U.S. government to purchase the private land is 20 million dollars.

Proposal #3

Establish a **large national park** by taking over three state parks plus 71,000 acres of private forest. Estimated cost to purchase land is 75 million dollars.

Or establish **no park** at this time.

Roles
Players will be given the roles of Senators and witnesses. Each witness is an important citizen with a special point of view on the park proposals. Each of the Senators gets a different role that shows his personality and how his supporters feel about the park issue.

Rules
1. The Senate Leader calls the hearing to order. He tells each witness when to testify. A witness has five minutes to make his presentation. Senators may then ask questions.

FIGURE 18.2
The Senate Chamber is the scene of debates on war, taxes, laws—and the redwood controversy.

2. After all witnesses have spoken, the Senate Leader asks each Senator to identify himself and state how he thinks he will vote. These preferences are recorded by the Leader on the Master Voting Sheet. (In real life, a Senate committee hears witnesses and reports to the rest of the Senate. The entire Senate then debates the issues and votes. For this game the two steps have been combined.)

3. Next, the Leader declares a 15 minute recess for discussion. Now, one Senator can try to get another Senator to change his vote.

FIGURE 18.3
Should these giant trees be preserved in a national park? How large should the park be?

4. Next, the Leader calls the roll, and each Senator casts his vote. These votes are also recorded on the Master Voting Sheet. The proposal which receives two-thirds of the votes wins. If no proposal receives enough votes, another discussion session is called. Go on with discussions and voting until one proposal wins.
5. Each Senator spins the Election Spinner to find out if he has been reelected. The chances for reelection are on each role. The data on reelection is then posted alongside the votes.
6. The class discusses what factors were crucial in settling "The Redwood Controversy."

Unit four

Man affects the environment

Would you be in favor of a highway being built that required your house to be torn down? Would you be in favor of having your neighborhood destroyed and made into a reservoir for fresh water? Do you think that inexpensive electricity is more important than clean lakes and streams? Highways, water systems, and power plants are only a few of the things that man has added to the environment.

Man's environmental changes are all around you. Some of them keep you warm, comfortable, and healthy. Other changes **pollute** (puh-LOOT) the environment and threaten your life. To pollute means to make the environment dirty and harmful to life. You probably hear a lot about pollution on radio and television and in the newspapers.

In Unit Four you will investigate how man affects the environment. You will learn to do these things:

a. Identify the ways that people (including you and your family) pollute your environment.
b. Predict some of the effects on the world if man continues to pollute.
c. How to change your behavior to make your community a healthier place.
d. Suggest ways to operate farms, homes, and factories that will be less destructive of your environment.

FIGURE 19.1
Everyone *thinks this sign is unfortunate—but who is responsible for it?*

Investigation

19 Who pollutes your environment?

Pollution problems are usually blamed on industries and the needs of large cities. But most citizens don't think they personally pollute their own environment, at least not seriously. Is this so? Or are you partly responsible for the "NO SWIMMING" sign at the beach? Do you add to the mess that makes your environment less beautiful and more deadly?

Problem 19-1

How do you affect your environment?

Think about the materials that you probably add to the water or soil. For example, when you wash a car, fertilize a lawn, take a bath, or even brush your teeth, you add new substances to the environment.

The soaps, toothpastes, and chemicals, along with human wastes, go into a drain or a sewer. Some cities may have sewage plants to partially treat the wastes. But whether the wastes are treated or not, most of the substances you add to the water cannot be removed. The sewer finally empties into a lake, river, or ocean. Some chemicals get into water in another way. Fertilizers and weed killers often sink down through the soil into underground water sources. This water eventually drains into streams and lakes. Do you think these chemicals affect our waters and the living things in them?

Who pollutes your environment?

FIGURE 19.2
Do these common household products add to the pollution of your environment? How many does your family use?

Do you think about pollution when someone spreads weed killer on your lawn, or when you run soapy bath water down the drain? Should you? This Investigation will help you find out how you affect your environment.

Gathering data

Think about an experiment you would like to do to discover how common household products affect living things. Figure 19.2 suggests some products you might like to test. You can probably think of others. Figure 19.3 gives a list of organisms you could study. It suggests some effects on these organisms to look for. Of course, you can select other organisms if you want. Plants and animals that live nearby are best. If possible, collect the organisms you decide to test.

Here are four suggestions to help you set up your investigation:

1. First select a product to test and an organism to test it on. Then find out about the normal behavior of your

FIGURE 19.3
Suggestions for investigating the effects of household products

Possible organisms collect them yourselves, if you can	Possible effects to study
algae, seaweed	growth rate, color change
leafy aquatic plants	growth rate, color change
hydra	rate at which it regrows parts
flatworms	rate at which it regrows parts
snails or periwinkles	growth rate, reproduction rate
daphnia, fairy shrimp	heartbeat rate
brine shrimps (available as eggs)	growth rate
crayfish	growth rate, breathing rate
larvae of aquatic insects, mayflies, dragonflies, mosquitos	rate of development
water striders	ability to stand on the surface
frog sperm and frog eggs	fertilization rate (number of eggs that start to hatch)
frog eggs, tadpoles	rate of development
young local fish—bluegills, sunfish, bullheads, minnows, catfish, mud minnows, top minnows	behavior, breathing rate
goldfish	breathing rate, behavior

organism and how to care for it. Find out where it lives and the kind of food it eats, if it is an animal. The library is a good place to start looking for information.

Suppose you decide to investigate how a chemical affects tadpole growth. How will you care for your tadpoles? Should you keep them in shallow water or deep water? What will you feed them and how often? Do they need to have plants in the water with them?

2. When you have learned about the life of your plant or animal, write out a plan for your investigation.

Decide on how much of the household product to use. Or how little. The amount of that substance in a bottle of lake water would probably be very small. You might

FIGURE 19.4
This pond is filled with a thick green scum of algae. What could cause so much plant growth?

test different amounts of the chemical on the growth of tadpoles. One container could test the effects of 1 ml of the chemical in 200 ml of water. Another container could test 0.1 ml of the chemical in 200 ml of water.

Tell your teacher if you need special instruments, like thermometers or a balance, for measuring changes in your organisms.

3. Run your experiment. Observe your organism very carefully. Look out for pollution effects you did not plan to test for. If it looks like the chemical is seriously harming your plant or animal, stop the experiment at once. Place the organism in clean water.
4. Report your results to the rest of the class.

While you are carrying out your experiment, also watch those of your classmates. Find out what questions they are investigating. See if you think their experiments are really asking the questions they say they are asking. Have all the experiments used controls?

Recording data

Keep accurate records of the information from your experiment. Include all the data, not just the information you believe is important. Organize your data in graphs or tables if you have many numbers.

Analyzing data

Write a report of your experiment which includes the question you were asking, your methods, your data, and your conclusions. Did the product you tested cause pollution?

You experimented with organisms alone in their containers. Think now about how changes in your organism might affect entire lakes, or rivers, or harbors. For example, did you or some of your classmates find that a substance killed plants or slowed plant growth? Perhaps a chemical you tested had the opposite result—it made plants grow very rapidly as in Figure 19.4. How might these changes affect an aquatic community?

Are there any ways that *you* can reduce pollution?

FIGURE 19.5

"Oh, it makes me so mad the way people pollute rivers!"

Who pollutes your environment?

Who pollutes your environment?

FIGURE 19.6
Do these activities add to the pollution in the environment?

Mastery Item 19-1

Family pollution

How much do you and your family contribute to the pollution of your environment? Your job is to try to answer this question. Look carefully all around your house, basement, and garage. Make a list of everything your family adds to the air, water, or soil. What suggestions can you make to reduce your family's contribution to local pollution?

Key

Discuss your list and your suggestions with your teacher.

Mastery Item 19-2

The dandelion problem

There is a small lake in the city park in Pine Falls, Ohio. In the hills around the lake are shady groves of trees and large, grassy picnic areas. Rain water runs off these hills into the lake. Many fish, frogs, turtles, and birds live in or near the lake. They also get their food from it.

Many people use the park in the summer for picnics and recreation. People who use the park sometimes complain about the large numbers of dandelions and other weeds in the picnic areas. These plants are considered to be less attractive than grass. Also, some of the broad-leafed plants grow best when there is bare earth around them. For many people, weeds make the areas much less desirable places to picnic.

The park department has been asked to solve "the weed problem."

Any of the following plans seem possible:

a. Every year apply a cheap, fast-working weed killer. This will destroy all of the undesirable plants. Grasses may fill in the spots where broad-leaved plants had been.
b. Ask everyone who uses the park to help dig up the weeds and plant more grass in the bare spots. This process, like spraying, would have to be repeated every two years.

FIGURE 19.7
Many people use and enjoy this park. Who is responsible for taking care of it?

c. Leave the picnic area pretty much as it is. There may be a good reason that the broad-leaved plants grow there instead of grass. Removing these plants by either method could have serious effects on the hills and lake. The weeds themselves are not a serious problem.

If you were the park manager, how would you rate these plans? What do you think would be the best choice?

List the plans in order, starting with the best one. Give a brief opinion of each plan.

Key

Choice c is probably the best plan because it does not damage the environment. Weeds such as dandelions may hold the bare soil in place where grasses won't grow. Choice b is a better idea than a, since it doesn't cause any pollution. Choice a could do serious damage to all the living things in the park.

Lake Erie gets used in many ways. Can the lake survive all these uses?

Investigation

20 Is Lake Erie dead?

Until a few years ago, Lake Erie meant little to most of the people in the United States. Today Lake Erie has become a symbol of water pollution. It is used as an example of the terrible things that can happen when people are careless about getting rid of wastes.

The newspapers and magazines say, "Lake Erie is dead," and "Lake Erie is a big open sewer." They may have exaggerated the facts. They may want to scare people into stopping the pollution. Most people who visit Lake Erie for the first time are surprised. They find that the water isn't "solid" with pollution, and fishermen still catch lots of perch and other fish.

Is Lake Erie really dead? Or are reports of Lake Erie's death only meant to scare? Is Lake Erie dying? Or is it healthier than ever before? Are all lakes bound to have the same things happen to them? Just what is happening in Lake Erie anyway?

Problem 20-1

What's wrong with Lake Erie?

Some people who have lived near Lake Erie all their lives say that the lake has gotten very bad. They say that it never used to stink. But it sometimes does now. The water never used to be scummy and foul-tasting. But it is now. These people may be right, but it's hard to be sure. Most people who talk about "the good old days" forget many of the bad things that happened back then.

Scientific data might help you to decide exactly what is happening to Lake Erie. Some quantitative data has been collected over the years. It is not nearly enough to give a complete answer, but it should help. You may find data that gives at least a partial answer to the question, "What's wrong with Lake Erie?"

Gathering data

On the following pages, you will find a "data bank." This bank contains many separate entries, collected from many sources. Each entry in the bank is data that has been gathered by scientific methods.

Choose one or more of the entries in the data bank. Carefully study the data in your entry. If your teacher gives you some other data on Lake Erie, search through it, too, for useful information.

Recording data

You don't need to write down any data, unless you have used a source besides the data bank. In that case, you should copy the data from your source.

Analyzing data

Try to find a pattern in the entry you study. Before you begin, look at this sample piece of data. Check to see that

Kind of organism		Requires high oxygen content in water	Animals per sq meter of bottom		
			1929	1930	1958
leeches		no	6	4	37
small worms		no	6	3	559
mayfly larvae		yes	312	515	49

you know how to analyze it. Does it tell of any changes in Lake Erie over the years?

The data shows that the numbers of organisms have changed greatly! In 1958 there were many more leeches and small worms than in earlier years. But there were fewer mayfly larvae in 1958. Why? Can you infer a possible cause for these changes by analyzing the table?

If you aren't sure how to read this sample table, ask your teacher for help.

Now select an entry in the data bank to study. You may want to rearrange or graph the data. That might help you analyze it. If you have a lot of trouble, tell your teacher. You probably saw earlier in this course that it is not always easy to find patterns in data. It takes some clever and careful detective work to find them. This is just what you are challenged to do in this Problem.

You will report any pattern you discovered to the whole class. Be sure you have evidence to support whatever pattern you found. Then your class may be able to put all the patterns together to find out what is wrong with Lake Erie.

It is possible that some entries in the data bank do *not* have a pattern in them. It is also possible that some entries do *not* contain useful information. Is your entry one of these? Be ready to give your reason if you think so.

The data bank

Entry 1

A **drainage basin** is an area that surrounds a lake or river. All the surface water in a basin drains into the lake or river. Within the drainage basin of the Great Lakes, shown in Figure 20.1, all rivers and streams flow toward the lakes. All their water finally ends up in the lakes. Outside the Great Lakes drainage basin, all rivers and streams flow away from the Great Lakes.

Entry 2

Figure 20.2a shows the shape of the bottom of Lake Erie. The curved lines connect points that are equally deep. The number on each line shows how deep it is along that line.

Figure 20.2b shows how Lake Erie would look if it could be sliced down the middle and viewed from the side. Notice

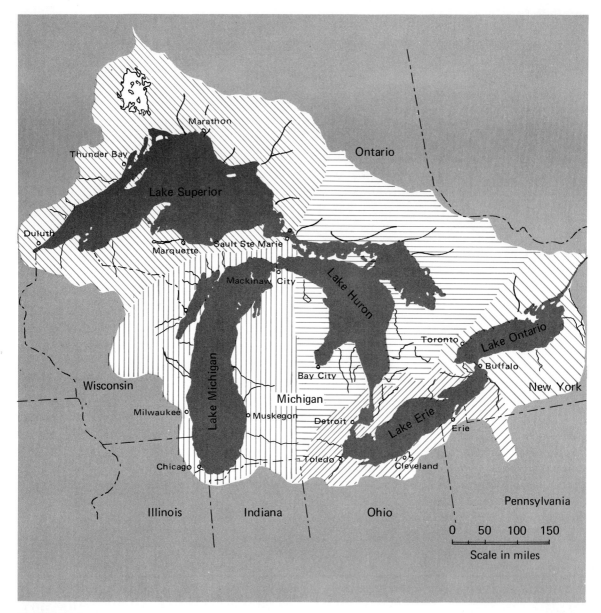

FIGURE 20.1
All the brooks, streams, and rivers within a lake's drainage basin flow into that lake. The drainage basin of all the Great Lakes covers a large area.

that in this drawing, the depth scale is *very* different from the length scale. The depth is in feet, and the length of the lake is in miles. How many parts does the lake seem to have? Each of these parts is called a lake basin because it looks something like a washbasin or sink.

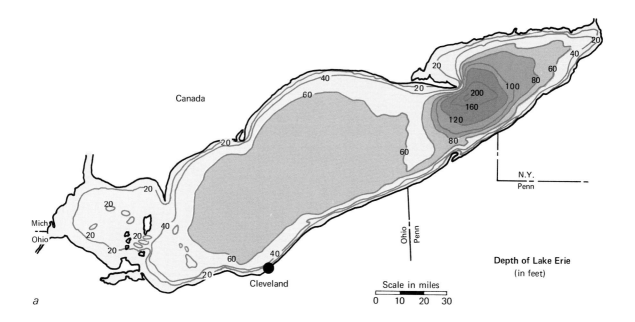

FIGURE 20.2
a. *This map is one way of showing the shape of the bottom of Lake Erie. The number on each line tells how deep the lake is at that place. The scale in miles is for the length of the lake.*
b. *This is another way to show the shape of the bottom of the lake. The vertical scale gives the depth in feet. The scale in miles is only for the length of the lake.*

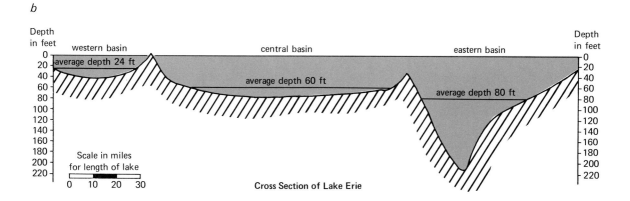

Is Lake Erie dead?

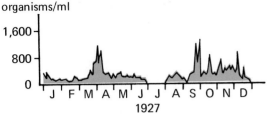
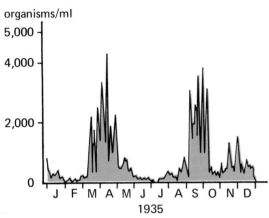
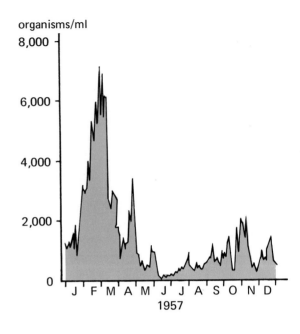
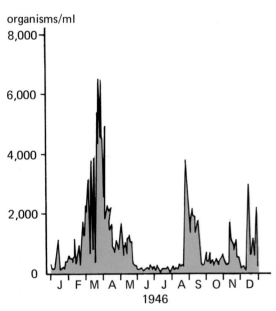
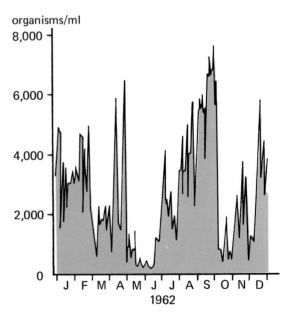

FIGURE 20.3
The graphs show the number of algae in Lake Erie at one of Cleveland's water intake pipes. Samples were taken daily. Did the pattern change between 1927 and 1962?

Entry 3

You may remember that algae are microscopic green plants that float in the water. Algae are the main food for some fish and other aquatic animals. Therefore, algae are the beginning of many food webs. There can be too many algae for two reasons. First, the organisms that eat it may be scarce. Second,

Is Lake Erie dead? 271

there may be chemicals in the water that act as fertilizers. Pollution can make algae grow very fast.

The people in Cleveland drink water from Lake Erie. One pipe that carries in the water extends about three miles out into Lake Erie. Every day a sample of water is taken near the pipe intake. A measured amount of the water is placed on a slide under a microscope. The number of algae is counted. Figure 20.3 shows some of the results.

Entry 4

Figure 20.4 shows some of the kinds of fish caught by commercial fishermen. It does not include fish caught for sport.

Most people find some kinds of fish tastier than others. They used to prefer sturgeon, pike, cisco, sauger, and whitefish. Lately, yellow perch, smelt, channel catfish, white bass, and carp are becoming more numerous.

FIGURE 20.4
This graph shows the catch of a few kinds of commercially important fish. Have the populations of desirable fish changed? Has the total production of fish in the lake changed?

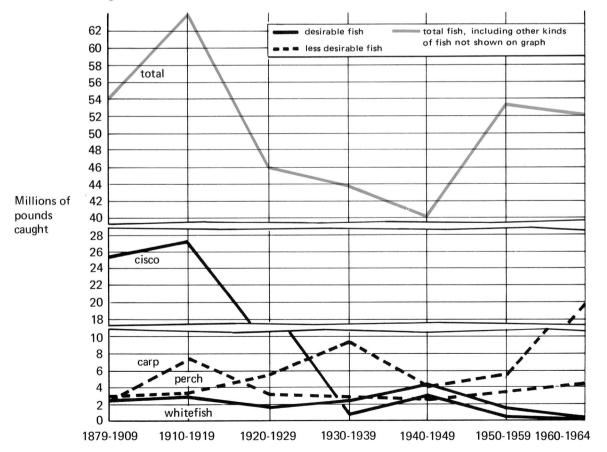

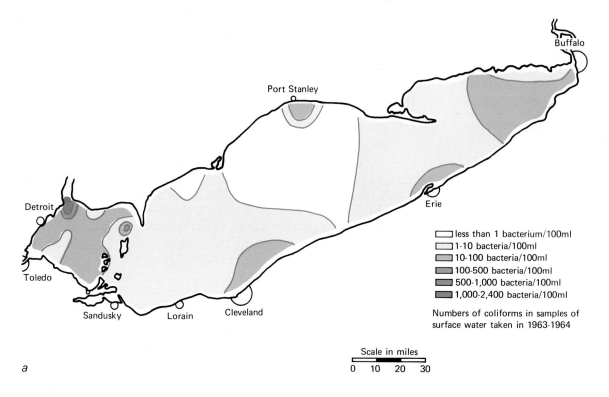

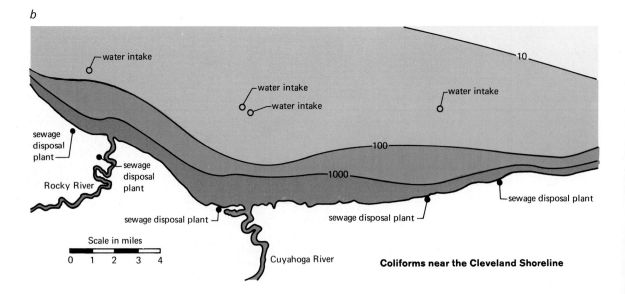

FIGURE 20.5

a. This map shows the number of coliform bacteria in Lake Erie. (*The number of bacteria close to shore isn't shown.*) Why do you think there are so few coliforms in the middle of the lake?

b. Here the coliform bacteria near Cleveland are shown. Do you think it is safe to swim in untreated water near Cleveland?

Entry 5

A kind of bacteria found in the intestines of humans is called **coliforms** (COE-lih-forms). They rarely cause any disease. But when they are found in water, it means that the water is polluted with sewage.

The more coliform bacteria present, the more the water is polluted. When there are many coliform bacteria, there are probably many disease-causing bacteria in the water, too. This is because the wastes of sick people are also being flushed into the water. Water is usually safe to swim in if it always has less than 1000 coliform bacteria per 100 ml of water.

Figure 20.5a does not show coliform numbers close to shore. Figure 20.5b shows the shoreline near Cleveland. The lines connect places with the same coliform counts. The beach below is at the eastern end of the lake.

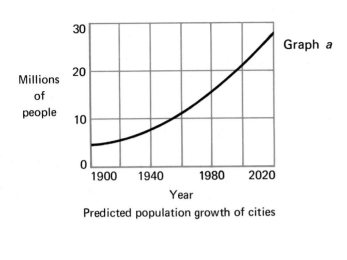

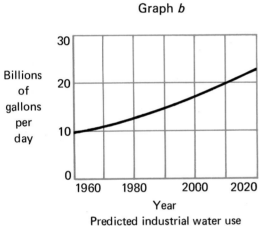

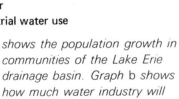

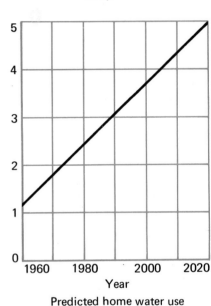

FIGURE 20.6
It is predicted that population and water use will change in the Lake Erie region. Graph a shows the population growth in communities of the Lake Erie drainage basin. Graph b shows how much water industry will use. Graph c shows how much water will be used in homes. What happens to all that water after it is used?

Entry 6
In using the data in Figure 20.6, be sure to notice that the time scale of Graph a is different from those of b and c. The "Lake Erie drainage basin" is shown in Entry 1.

Entry 7
All water in nature contains living organisms and dead matter. The dead matter can be dead organisms, human wastes—

FIGURE 20.7

Chemical oxygen demand in Lake Erie (mg/liter)

Cruise number	Date	Western basin			Central basin			Eastern basin		
		high	low	avg.	high	low	avg.	high	low	avg.
52	10/63	10.6	3.5	6.5	8.6	3.1	6.4	20.9	4.7	6.9
57 and 58	5/64	13.1	7.5	10.7	16.0	3.6	8.4	27.0	5.7	8.8
61 and 62	6/64	28.0	4.2	12.3	6.0	5.0	5.5	7.0	6.0	6.1
66	8/64				9.0	7.0	8.1	13.0	6.0	8.0
67	9/64	29.0	1.1	12.0						
average				10.4			7.1			7.5

Source: *Lake Erie Environmental Summary*, United States Department of the Interior, Federal Water Pollution Control Administration, 1968.

anything that was once alive. Polluted water often contains a lot of dead matter. All the dead things in the water eventually decay. Decay uses up oxygen. This may leave very little oxygen dissolved in the water. Then some organisms may not have enough oxygen to live. In fact, the dissolved oxygen may be all used up.

Chemists have a test to measure how much oxygen the decaying matter in the water will use up. The amount of oxygen that will be used up is called the **chemical oxygen demand,** or COD. A water sample with a COD of 10 mg/liter will use up twice as much oxygen as a sample with a COD of 5 mg/liter.

Usually water can hold up to about 20 mg/liter of dissolved oxygen. Imagine that a sample of water contains 20 mg/liter of dissolved oxygen. Say its COD is 25 mg/liter. How much dissolved oxygen would you expect the water to contain after a week or so?

The COD's in Figure 20.7 were collected during cruises of a research boat. Not all of the measurements from each cruise are given. Only the highest, lowest, and average measurements are shown.

Notice that Lake Erie is divided into three areas, or basins. Figure 20.2b shows where these basins are.

Entry 8

The data in this entry shows how many millions of gallons of waste water flow into Lake Erie every day

Two kinds of sewage treatment plants pour water into the lake. **Primary treatment plants** remove only the floating material in sewage. Anything dissolved in sewage or too small to filter, like bacteria, goes into the lake.

Some places also have **secondary sewage plants.** Secondary treatment plants make the waste water from primary sewage plants much cleaner. They use microbes to break down many undesirable substances in the sewage. You may have observed similar actions of microbes in Investigation 15.

Many industries also pour waste water into the lake. These include electric power plants, steel and paper mills, and oil and chemical companies.

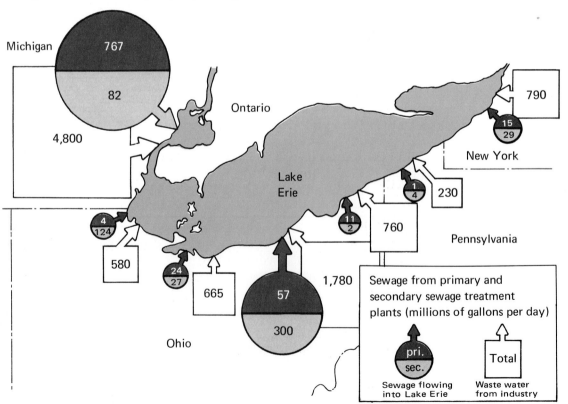

FIGURE 20.8
The amount of waste water flowing into Lake Erie every day is measured in millions of gallons. Who is responsible for polluting Lake Erie?

Is Lake Erie dead?

Entry 9

Figure 20.9 shows the number of animals living on the bottom of western Lake Erie in three different years. Scientists survey populations of these organisms by scooping up all of the mud, sand, or rocks in a measured area. They dig down about one inch. The sample is placed in a container. The animal organisms are removed, classified, and counted. They can all be seen without a microscope.

FIGURE 20.9

Bottom-dwelling organisms in western Lake Erie. **Larvae** *(LAR-vee) are young organisms. Both young and adult mayflies are a very important source of food for lake fish. Small worms and fingernail clams aren't nearly as good for fish to eat. Leeches are worms that may be blood-suckers. Midges are small, often pesty, flies. Most desirable kinds of fish, such as trout, require a lot of oxygen.*

Kind of organism		Requires high oxygen content in water	Animals per sq meter of bottom		
			1929	1930	1958
leeches		no	6	4	37
small worms		no	6	3	559
mayfly larvae	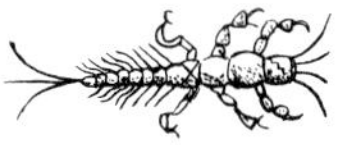	yes	312	515	49
midge fly larvae	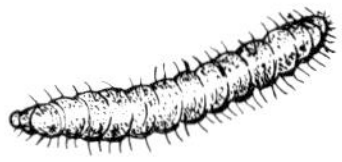	no	81	22	257
snails		yes	12	24	7
fingernail clams		no	16	8	55
caddis fly larvae		yes	20	1	3

Source: Ohio Journal of Science, 1960.

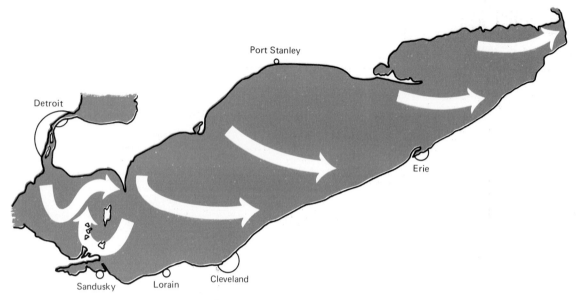

FIGURE 20.10
The flow of water at the surface of Lake Erie during the summer can be shown on a map. Does the map help you to explain any of the other data bank entries?

Entry 10

Notice that the arrows on the maps show the usual direction of water flow in the summer. The water does not flow in those directions all the time. Also, the data doesn't tell how fast the water flows.

Entry 11

Polluted lakes usually have a heavy growth of algae. This often looks like a green scum on the water. The many live algae and the dead and decaying algae give off a rotten, fishy smell.

The exact cause of heavy growths of algae in Lake Erie has been discussed and argued for years. Scientists don't yet agree on the cause. Man has polluted Lake Erie with many substances. Two chemicals are usually blamed for the increased growth of algae: **nitrogen** (NY-troh-jen) compounds and **phosphorus** (FOSS-fur-us) compounds. The data in Figure 20.11 is from a laboratory experiment. It was designed to test the importance of these compounds.

Samples of water from Lake Erie were collected in November, January, May, and July. They were combined. Test containers were partly filled with 200 ml of the water. Then 4 ml

FIGURE 20.11

Growth of algae in lake water. Which container of water grew the most algae?

	Sample	Total phosphorus (mg/liter)	Total nitrogen (mg/liter)	Color of filters 10 days	30 days
a	lake water alone	0.01	0.42		
b	lake water plus 4 ml filtered raw sewage	0.07	1.18	●	●
c	lake water plus 4 ml treated sewage from secondary sewage plant	0.14	1.18	●	●
d	lake water plus 4 ml sewage *plus* special removal of phosphorus compounds	0.02	1.13		○
e	same as **d** *except* phosphorus compounds added back after treatment	0.14	1.13	●	●

Source: Adapted from Vallentyne, J. R. et al., "Phosphorus and the Control of Eutrophication in the Lower Great Lakes" (in publication) and *Lake Erie Environmental Summary*, United States Department of the Interior, Federal Water Pollution Control Administration, 1968.

of another liquid was added. These second liquids are described under Figure 20.11. The figure also shows the amounts of phosphorus and nitrogen in each container. There is as much phosphorus in **e** as in **c**.

Algae were grown in the test containers in the light at 22°C. One set was grown for 10 days and another set for 30 days.

The contents of the flask were filtered through plain white filters at the end of the growth period. The filters caught the algae. If only a few algae grew in the water, the filter was light-colored. If many grew, the filter was a bright green.

Is Lake Erie dead?

Entry 12

Homes and industries around Lake Erie produce a lot of soapy water. Much of it ends up in the lake. Detergents contain phosphorus compounds called **phosphates** (FOSS-fayts). Detergent makers may change the amount of phosphates in a particular detergent from time to time. Therefore, by now there may be more or fewer phosphates than listed in Figure 20.12.

FIGURE 20.12
Phosphates in major detergents. Following is a list of the percentages of phosphates in major detergents, as compiled in a study by Limnetics, Inc., a Milwaukee consulting concern.

Detergent	Manufacturer	Per cent phosphate
Axion	Colgate-Palmolive	43.7
Biz	Procter & Gamble	40.4
Bio-Ad	Colgate-Palmolive	35.5
Salvo	Procter & Gamble	35.3
Oxydol	Procter & Gamble	30.7
Tide	Procter & Gamble	30.6
Bold	Procter & Gamble	30.2
Ajax Laundry	Colgate-Palmolive	28.2
Punch	Colgate-Palmolive	25.8
Drive	Lever Brothers	25.3
Dreft	Procter & Gamble	24.5
Gain	Procter & Gamble	24.4
Duz	Procter & Gamble	23.1
Bonus	Procter & Gamble	22.3
Breeze	Lever Brothers	22.2
Cheer	Procter & Gamble	22.0
Fab	Colgate-Palmolive	21.6
Cold Power	Colgate-Palmolive	19.9
Cold Water All	Lever Brothers	9.8
Wisk	Lever Brothers	7.6
Diaper Pure	Boyle-Midwest, Inc.	5.0
Trend	Purex Corporation	1.4

Source: New York Times, December 14, 1969.

Is Lake Erie dead?

Problem 20-2

Can Lake Erie be helped?

Many people want to save Lake Erie. But few people know what's wrong with it. They just know "It's bad." On the other hand, you really know a lot about Lake Erie. Now that you've done Problem 20-1, you should be able to suggest how Lake Erie can be helped.

Some of the entries in the data bank that were not useful in Problem 20-1 may be useful in this Problem.

Gathering data
Use the data bank or any other source of information to get ideas. Your goal is to come up with at least one plan which would improve Lake Erie and could really be put into action.

Recording data
Make notes if you use information not in the data bank. It is important to note the exact source of your data. If your ideas come from the data bank, you might note which entry or entries they came from.

Analyzing data
Write a few sentences to describe your plan to improve Lake Erie. Be ready to give reasons for your plan in a class discussion. You should also try to decide if other suggested plans would be helpful.

Mastery Item 20-1

Which lake is more polluted?

Which lake do you think is more unpleasant to be near—Lake Erie or Lake Ontario? Figure 20.13 gives some data on each lake. Use only the data provided. If you don't know what a characteristic means, find the data bank entry that describes it. Try to decide which of these two lakes is probably more polluted. Give the reasons for your choice.

Is Lake Erie dead?

FIGURE 20.13
Some characteristics of Lake Erie and Lake Ontario

Characteristic	Lake Erie	Lake Ontario
Drainage basin (sq mi)	33,000	35,000
Average depth (ft)	58	283
Annual commercial fish catch (million lbs)	48	2
Average surface temperature (°C)	10	9
Dissolved oxygen (mg/liter)*	39	37
Total nitrogen (mg/liter)*	0.6	0.2
Total phosphorus (mg/liter)*	0.03	0.01
Total chlorine (mg/liter)*	25	35
Chemical oxygen demand* COD (mg/liter)	9.5	6.6
Total coliform bacteria* (organisms/100 ml)	12	8
Algae* (organisms/100 ml)	12,000	4,000

*Average for the entire lake for one year

Key

The data indicates that Lake Erie is more polluted than Lake Ontario. You might have given any of the following reasons:
1. higher chemical oxygen demand
2. more coliform bacteria
3. more phosphorus and algae

Mastery Item 20-2

Is Crystal Lake dying?

Imagine that you live in a city near Crystal Lake. This large lake is a popular recreational spot. The local newspaper one day carried a front page story with the headline: "IS CRYSTAL LAKE DYING?"

Here is a list of questions about Crystal Lake. The newspaper article answered some of them. Most of the questions were not answered (or asked). Which questions do you think would help you decide how polluted Crystal Lake is?

Mark each question either **Y** (yes, the answer would be helpful) or **N** (no, the answer probably would not help).

1. How thick was the ice last winter?
2. Are there more dissolved sulfur compounds now than a year ago?
3. Is there enough oxygen dissolved in the lake water to support many kinds of organisms?
4. Has the number of water-skiers changed in the past 10 years?
5. How big was the biggest fish caught last summer?
6. Does untreated sewage flow into the lake?
7. Are there more algae in the water this week than there were last week?
8. Have some kinds of fish become scarce during the past 10 or 20 years? Have some other kinds become plentiful?
9. Are there any bacteria in the lake now?
10. Are there any phosphorus compounds in the lake water?
11. From what direction does the wind usually blow in August and September?
12. Has the amount of algae in the lake during June and July changed in the past 10 years?
13. Has the number of sport fishermen visiting the lake changed during the past 10 years?
14. Have the kinds of worms and other small animals living on the bottom of the lake changed during the past 10 years?
15. Has the horsepower of motorboats on the lake increased much in the past 10 years?
16. What is the concentration of coliform bacteria in the lake now?
17. Can modern water treatment methods make the lake water safe to drink?
18. Are there any snakes living in the lake or along its shores?
19. Does the lake water ever look extremely green and scummy for a few days?
20. Have property values around the lake gone up recently?

Is Lake Erie dead?

Key

The best answers to the questions probably are:

1—N	5—N	9—N	13—N	17—N
2—Y	6—Y	10—Y	14—Y	18—N
3—Y	7—N	11—N	15—N	19—Y
4—N	8—Y	12—Y	16—Y	20—N

If you have other answers, discuss them in class.

Mastery Item 20-3

Can Crystal Lake be saved?

Imagine that you are the Commissioner of Water Pollution Control in Crystal City. You have been shown three plans for reducing the pollution in Crystal Lake. You can decide which plan to carry out. Each of the plans will cost the same amount of money and take the same amount of time to carry out.

Decide which plan is best. Write out your decision and your reasons for making it.

PLAN A

Build primary and secondary sewage treatment plants for all the cities in the drainage basin.

In each plant set up a process to change 90 per cent of all phosphorus compounds in the sewage into chlorine compounds.

PLAN B

Build primary sewage treatment plants for all the cities in the drainage basin.

Build a pipeline to carry all the waste water from these sewage treatment plants to a large river which flows into the ocean.

PLAN C

Build primary and secondary sewage treatment plants for all the cities in the drainage basin.

Ban the use of all detergents that contain a lot of phosphorus compounds.

Stop using certain agricultural fertilizers in the drainage basin. Fertilizers contain large amounts of nitrogen and phosphorus compounds. These compounds may wash into streams that flow into Crystal Lake. Use only fertilizers that don't wash away so easily.

Key

Plan A You probably don't know whether chlorine compounds are harmful to living things. Chlorine compounds may cause pollution, too.

Plan B This plan just sends the undesirable substance somewhere else, where they will become a big problem for lots of other people and aquatic organisms.

Plan C This is probably the best plan. If Crystal City is going to stop polluting the lake, its citizens must stop using the products that cause pollution.

FIGURE 21.1
above, *An atomic power plant pours heated water into a river.*
right, *Water cools hot metal in a steel factory.*

Investigation

21 What is the price of progress?

What do you think about when you see and smell dirty water? Probably everyone likes clean water. How, then, did most rivers, lakes, and ocean harbors get so polluted? Why do people pollute water? This Investigation will help you answer these questions.

You already know that clean water is polluted by household items, chemicals, sewage, and detergents. You can probably think of other pollutants, too.

You may not have known that cool, clean water can be polluted by dumping hot water into it. This is called **thermal** (THUR-mul) pollution. Figure 21.1 shows some common sources of thermal pollution.

Thermal pollution keeps company with other kinds of pollution. The same waterfront factories and power plants often dump out chemicals, oil, acids, *and* hot water. Therefore, when you study thermal pollution, you will also learn why other kinds of water pollution have taken place.

There are two parts to this Investigation. The first Problem is a game for groups of players. You will take an imaginary trip back to the 19th Century to begin the game. There you will invest your money in growing industries and try to get rich. In the second Problem, you may be an investor again, or a politician, or a concerned citizen. Your job will be to decide whether to allow a new source of thermal pollution to be built in your town.

The price of progress?

Problem 21-1

How did the rivers get hot?—The Thermal Pollution Game

The game takes place in Central City, in the northeastern United States. The game begins in 1885 and runs years into the future. During this time, the city grows—population increases, big new factories are built, suburbs are made out of woods and farms, and everyone buys a car. Four steel mills are built on one river bank alone. The largest is one-half mile long. Central City is industrial America.

There are five time periods to the game. They stand for different stages of Central City's growth. In each period, you can make a fortune and affect the rivers.

Your main purpose in this game is to become richer than the other players in your group. You earn money by investing in industries and other businesses. You will see what happens to a water resource when everyone tries to make a lot of money.

Read the following description of Central City from 1885 to 1905; then start the play.

Central City 1885–1905

Central City is a fast-growing industrial city. Its population increases from 300,000 in 1885 to 500,000 by 1905. Many of the new people are workers from Europe. They swarm into the city with their large families. They are eager to work in the factories, even though wages are low.

Most of the workers rent houses near the factories and the smoke. Their houses are heated by coal fires and lit by gas or kerosene lamps. Their neighborhoods are dirty, noisy, and crowded. Families with more money move outside of town.

The railroad is the king of transportation. It has lately become possible to travel from New York to California by rail. Horse-drawn buggies and wagons are the main vehicles in the city. But nearly everyone walks to work.

Most of the factories are built near the rivers. The rivers serve as a dump for industrial wastes. Factories also use the

FIGURE 21.2
Central City in 1885 is just starting to grow rapidly.

river water to power steam engines and to cool machines and hot metals. The rivers are also used to transport barges of coal, oil, and iron ore. Finally, the rivers supply drinking water.

Materials
game board
a die
chips or play money
tracing paper
2 containers of water
thermometer

The price of progress?

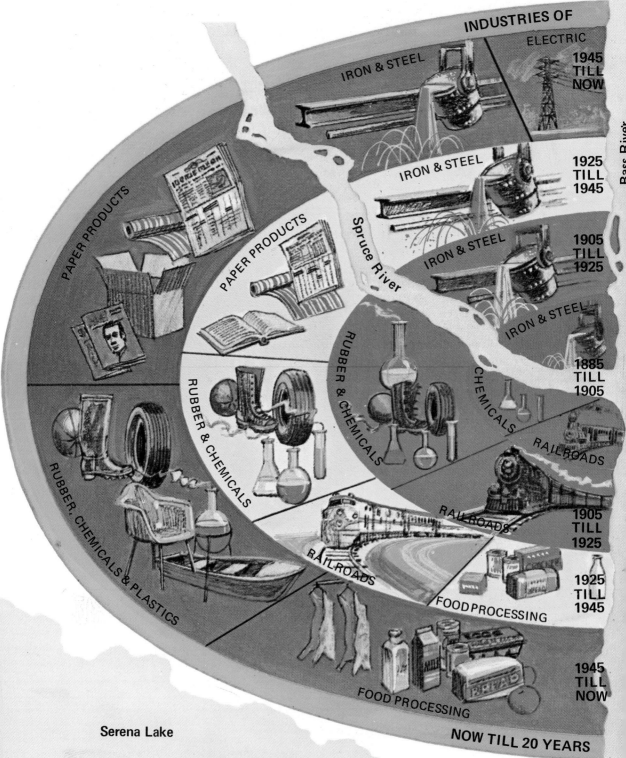

The price of progress?

FIGURE 21.3
Use this board to play The Thermal Pollution Game.

The price of progress?

Playing rules

1. Each player tosses a die once. The number that comes up tells how many thousands of dollars he gets to start the game. White chips are worth $1,000 each; red chips, $5,000; and blue chips, $10,000. One player serves as banker.
2. The first round of play starts in the center of Central City. Figure 21.3 is the thermal pollution game board you will be using. The center area is marked 1885–1905. Each player can invest any part of his money in any of the industries there.

    When you invest money in a company, it means you own a part of the company. If business is good, you will get back more money than you put in.
3. Each player chooses one of these symbols:

    In each symbol he will record how much he invests. He uses the same symbol for the whole game. The investment records for all players are made on a sheet of tracing paper laid over one game board.

    Imagine a player chose the triangle. He then decides to invest $4,000 in the oil industry. Therefore, on the tracing paper, over Oil Products, he writes:

    △4
4. After all players record their investments, each tosses the die once more. The number that comes up is used with the Income and Temperature Index for 1885–1905. This tells him how much money he earns by 1905.

    To find the income, he multiplies his investment in a business by the proper income increase. For example, a player has $4,000 invested in Oil Products, and $1,000 in Chemicals. His die comes up a 4. He multiplies each investment by the proper income increase (2 × $4,000 and 1 × $1,000) then he collects the total amount ($9,000) from the banker.
5. The Index also tells how much heat each industry and its factories added to the river waters. The player's investment in Oil Products increased the temperature by 0.25°F, and the chemical investment didn't increase the temper-

ature at all. Each player records the total number of degrees that his investments heated the rivers, for example,

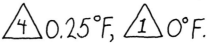

6. Now it's time to add up the total number of degrees the rivers have been heated. The normal river temperature is 70°F in the summer. Add to that the total heat increase. This new temperature tells how warm the water has become. Record the new water temperature over the year 1905 on the tracing paper.

FIGURE 21.4

Income and Temperature index: 1885–1905.

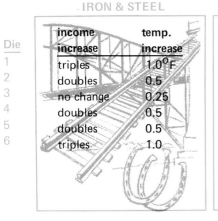

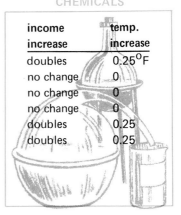

IRON & STEEL

Die	income increase	temp. increase
1	triples	1.0°F
2	doubles	0.5
3	no change	0.25
4	doubles	0.5
5	doubles	0.5
6	triples	1.0

OIL PRODUCTS

income increase	temp. increase
no change	0°F
doubles	0.25
doubles	0.25
doubles	0.25
doubles	0.25
no change	0

CHEMICALS

income increase	temp. increase
doubles	0.25°F
no change	0
no change	0
no change	0
doubles	0.25
doubles	0.25

RAILROADS

Die	income increase	temp. increase
1	triples	1.0°F
2	doubles	0.5
3	doubles	0.5
4	no change	0.25
5	doubles	0.5
6	triples	1.0

BUGGIES & WAGONS

income increase	temp. increase
no change	0°F
doubles	0
doubles	0
no change	0
no change	0
doubles	0

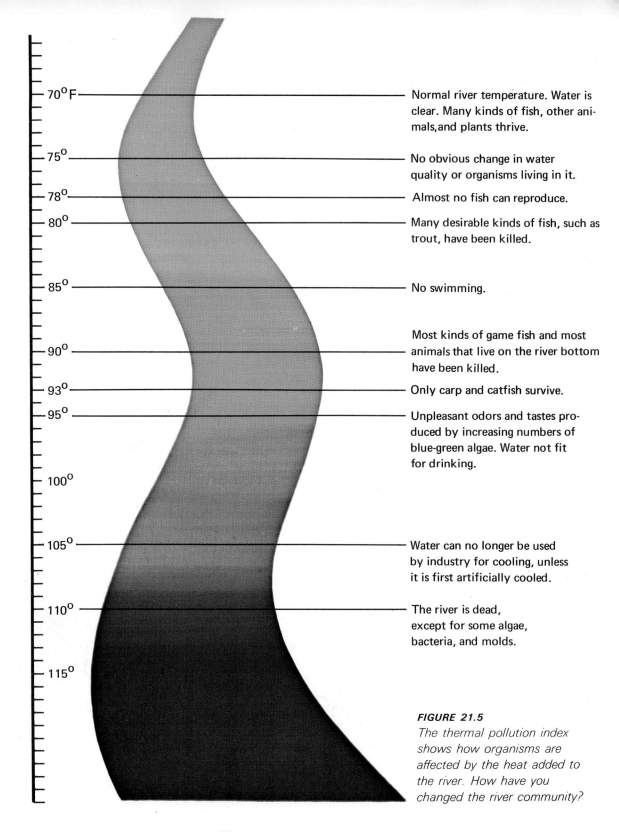

FIGURE 21.5
The thermal pollution index shows how organisms are affected by the heat added to the river. How have you changed the river community?

7. All players examine Figure 21.5 to see how the temperature increase affects the rivers of Central City. Any player concerned about the effects of thermal pollution can try to do something about it. In the next round he can invest money in industries that he thinks won't add a lot of heat to the water. (For this Problem, you are concentrating on waste heat, not other pollutants that are entering the river.)
8. Start the next round of play by reading about the growth of Central City.

Central City 1905–1925

The Central City area becomes one of the nation's largest industrial centers. New industries and old ones grow rapidly. A web of railroad tracks is built to connect American cities. The automobile industry grows up. Henry Ford invents his Model T car in 1908 and sells two million of them in 1923 (at $260 apiece).

Electricity becomes widely used by homes and factories. Huge amounts of fuel are used to provide power for manufacturing. The city electric plants burn coal to make steam to turn generators to make electricity. Excess hot water is discharged into the rivers.

FIGURE 21.6
Central City in 1905 has become a large industrial center. The mills along the river employ thousands of people.

The price of progress?

IRON & STEEL

Die	income increase	temp. increase
1	doubles	2.0°F
2	triples	3.0
3	doubles	2.0
4	triples	3.0
5	doubles	2.0
6	triples	3.0

OIL PRODUCTS

Die	income increase	temp. increase
1	no change	0.5°F
2	doubles	1.0
3	triples	1.5
4	doubles	1.0
5	no change	0.5
6	doubles	1.0

RAILROADS

Die	income increase	temp. increase
1	doubles	3.0°F
2	triplets	3.0
3	triples	4.0
4	triples	4.0
5	doubles	3.0
6	doubles	3.0

RUBBER & CHEMICALS

Die	income increase	temp. increase
1	triples	2.0°F
2	triples	2.0
3	no change	0.5
4	doubles	1.0
5	doubles	1.0
6	doubles	1.0

AUTOMOBILES

Die	income increase	temp. increase
1	lost all	1.0°F
2	no change	1.0
3	doubles	2.0
4	triples	2.5
5	doubles	2.0
6	no change	1.0

ELECTRICAL POWER

Die	income increase	temp. increase
1	doubles	3.0°F
2	doubles	3.0
3	triples	4.0
4	triples	4.0
5	triples	4.0
6	doubles	3.0

BUGGIES & WAGONS

Die	income increase	temp. increase
1	no change	0°F
2	lost all	0
3	lost all	0
4	no change	0
5	doubles	0
6	no change	0

FIGURE 21.7
Income and Temperature index: 1905–1925

The price of progress?

The population jumps from 500,000 to 1,000,000 people. They, too, demand more power. Throughout these years, people from Europe continue to flock to the city. It has a good reputation. If a man works hard, he can earn a good living there. Some people with cars move to the suburbs. They want to escape the environment of the industrial center. But most people cannot or do not want to move.

The city becomes dirtier and dirtier. The rivers downstream from the industries are full of acids, chemicals, and old tires.

9. Each player reinvests his money. Follow rules 3–7 again. *Electrical Power and Iron and Steel must each receive investments from at least one player.* Use the new Income and Temperature Index for 1905–1925.
10. Play continues the same way for the next two periods, 1925–1945 and 1945–Now. The city is described during each period. Each period also has its own Income and Temperature Index.

Central City 1925–1945

During this era, Central City continues to expand. More people use the automobile and improved public transportation to get from home to work. More bridges are built across the rivers. The population grows to 1,200,000.

FIGURE 21.8
Central City in 1945 is the largest and dirtiest industrial center in the country. Notice how the industries along the river have grown.

IRON & STEEL

Die	income increase	temp. increase
1	doubles	4.0°F
2	no change	2.0
3	doubles	4.0
4	doubles	4.0
5	doubles	4.0
6	triples	6.0

OIL PRODUCTS

Die	income increase	temp. increase
1	lost all	0°F
2	doubles	1.0
3	doubles	1.0
4	doubles	1.0
5	doubles	1.0
6	no change	0.5

RAILROADS

Die	income increase	temp. increase
1	no change	1.0°F
2	doubles	2.0
3	no change	1.0
4	lost all	0
5	doubles	1.0
6	no change	2.0

RUBBER & CHEMICALS

Die	income increase	temp. increase
1	doubles	2.0°F
2	triples	3.0
3	no change	1.0
4	no change	1.0
5	lost all	0
6	doubles	2.0

MOTOR VEHICLES: CARS, TRUCKS, BUSES

Die	income increase	temp. increase
1	doubles	2.0°F
2	no change	1.0
3	lost all	0
4	no change	1.0
5	doubles	2.0
6	doubles	2.0

ELECTRICAL POWER

Die	income increase	temp. increase
1	doubles	6.0°F
2	no change	3.0
3	triples	9.0
4	doubles	6.0
5	doubles	6.0
6	triples	9.0

PAPER PRODUCTS

Die	income increase	temp. increase
1	no change	0°F
2	doubles	0.5
3	doubles	0.5
4	no change	0
5	doubles	0.5
6	lost all	0

LAND DEVELOPMENT

Die	income increase	temp. increase
1	triples	3.0°F
2	doubles	2.0
3	no change	1.0
4	doubles	2.0
5	doubles	2.0
6	doubles	2.0

FOOD PROCESSING

Die	income increase	temp. increase
1	doubles	0.5°F
2	no change	0
3	no change	0
4	no change	0
5	lost all	0
6	doubles	0.5

FIGURE 21.9
Income and Temperature index: 1925–1945

The price of progress?

Suddenly the economy of the nation and the world collapses. There is a Great Depression. Business expansion is slowed for a decade. Fortunes are lost in the stock market, and businesses and banks fail. Thousands of workers lose their jobs.

Then there is a slow recovery that lasts until 1940 and the beginning of World War II. During the war, Central City becomes the center of a rich industrial empire. Workers make new products that their grandfathers never dreamed of. The work week is reduced from 60 to 40 hours for most workers, and they get more pay.

The water, both upstream and downstream, is terrible. For the first time Central City has to rely on Serena Lake for its drinking water. The taxpayers must pay for a pumping station and a pipeline from the lake to the city.

Central City 1945–Now

During this period the city itself loses population. People still migrate from farms and small towns to find jobs. But the rush to the suburbs is on. Farmland is turned into suburbs. Suburbs grow from farm villages into cities. Some of these cities are very large. The population of the Central City area passes 2,000,000.

FIGURE 21.10

Central City now has many new office buildings for its industries. Industries are still growing along the river and polluting it.

The price of progress?

The city is known as "the crossroads of industry." More than half the people in the nation live within 500 miles. They are a tremendous market for the city's products. Progress, however, has a price. Sometimes the street lights are turned on at noon. The air is that smoky. The rivers are equally polluted. Citizens of the city and its suburbs must begin to find solutions to the pollution problem, or the area will no longer be fit to live in.

In the meantime, Serena Lake has become a very popular resort. Many people come from other parts of the state and from nearby states to camp, fish, swim, and water-ski. In Problem 21-2, you will decide if Serena Lake will be polluted.

11. Do not play the last round (Now—20 years in the future) at this time. You will have a chance to play it later. To determine the winner, each player adds up his wealth. You can find the winner in your group and your class.

Post-game discussion

Discuss the game with your group. You can start by preparing one container of water at 70°F. That is the temperature of the rivers in 1885. Then mix hot and cold water in another container. Match the temperatures of the river water at the end of each time period. Compare the water in both containers by *carefully* dipping in your hands.

Here are some questions to discuss. You will probably think of others, too.

a. Where did most of the heat in the river seem to come from?
b. Can an industry or business cause thermal pollution indirectly?
c. Are there any businesses that don't cause much thermal pollution?
d. Can you become wealthy and still have an unpolluted river?
e. Are there any winners of the game when it comes to water quality?

Has your discussion given you ideas about how to earn money without polluting the river a lot? Play the game again and test your ideas.

The price of progress?

IRON & STEEL

Die	income increase	temp. increase
1	triples	9.0°F
2	doubles	6.0
3	triples	9.0
4	doubles	6.0
5	triples	9.0
6	triples	9.0

OIL PRODUCTS

Die	income increase	temp. increase
1	doubles	3.0°F
2	doubles	3.0
3	triples	4.0
4	triples	4.0
5	triples	4.0
6	triples	4.0

RUBBER, CHEMICALS & PLASTICS

Die	income increase	temp. increase
1	triples	9.0°F
2	triples	9.0
3	triples	9.0
4	doubles	6.0
5	doubles	6.0
6	doubles	6.0

MOTOR VEHICLES & AIRCRAFT

Die	income increase	temp. increase
1	no change	2.0°F
2	doubles	4.0
3	triples	6.0
4	triples	6.0
5	triples	6.0
6	doubles	4.0

ELECTRICAL POWER

Die	income increase	temp. increase
1	triples	12.0°F
2	triples	12.0
3	doubles	8.0
4	doubles	8.0
5	triples	12.0
6	triples	12.0

PAPER PRODUCTS

Die	income increase	temp. increase
1	triples	3.0°F
2	doubles	2.0
3	triples	3.0
4	doubles	2.0
5	triples	3.0
6	doubles	2.0

LAND DEVELOPMENT

Die	income increase	temp. increase
1	triples	9.0°F
2	triples	9.0
3	triples	9.0
4	doubles	6.0
5	four times	12.0
6	four times	12.0

FOOD PROCESSING

Die	income increase	temp. increase
1	doubles	2.0°F
2	no change	1.0
3	triples	3.0
4	triples	3.0
5	doubles	2.0
6	doubles	2.0

ELECTRONICS: RADIO, TV, HI-FI, COMPUTERS

Die	income increase	temp. increase
1	triples	2.0°F
2	no change	1.0
3	four times	2.0
4	five times	3.0
5	doubles	1.0
6	triples	2.0

FIGURE 21.11

Income and Temperature index: 1945–now

The price of progress?

Problem 21-2

The atomic power plant controversy

In Problem 21-1 you saw what has happened to the quality of water in some rivers. That is history. What will happen to other water resources now and in the future? Who will care about the quality of water and who will decide what our lakes and rivers are used for? To be more specific, what will happen to the water in Serena Lake?

There is a problem that exists in Central City today and may become even more serious in the future. The city needs more electricity. But heat from more power stations may damage the environment.

The Wattson Electric Company is planning to build a huge power station on Serena Lake. Figure 21.12 shows what the station will look like. Atomic energy will supply the heat for making electricity. Waste heat will be given off as hot water. The company doesn't plan to cool the water before piping it into the lake.

FIGURE 21.12
This is what the Wattson Atomic Power Station will look like when it is finished. Do you think it should be built as planned?

The price of progress?

Before the company can begin operations, it must get permission to release heat into the lake. The State Health Board is responsible for granting the permit.

Some citizens from Central City do not want the state Health Board to grant the permit. Other groups and organizations support the power company. A hearing will be called by the State Health Board to decide the matter. There are two issues.

1. *Should the Wattson Electric Company be permitted to go ahead with its present plans for discharging hot water into Serena Lake?*
2. Suppose the company is not allowed to add hot water to the lake. *There are other ways to get rid of the waste heat. Will any of these methods be allowed?*

You will be playing a role, just as in The Redwood Controversy. Your goal is to convince your opponents and the Health Board to support your position. Use all the data and arguments you can. Use all your powers of persuasion. If you are elected to the Health Board, you get to judge the case.

Materials

descriptions of peoples' roles, on pages 304–313
Data Bank, on pages 313–323

How to play

1. The class elects players to be on the State Health Board. Your teacher will tell you how many Board members to elect and how the voting will take place.
2. If you were not elected, choose one of the remaining roles listed on the chalkboard. Then get together with your classmates who chose the same role. Stay together as a group for the rest of this Investigation.
3. Your group should plan its strategy. Find the description of your role on pages 304–313. It tells you who you are and what you stand for. Read the role and discuss it within your group. You will also need to analyze and discuss the data on pages 313–323. *Be sure your group can use this data to support your position.*
4. The State Health Board calls the hearing to determine what kind of permit, if any, will be granted. You will present your views at the hearing.

5. Choose one member of your group to introduce your group and present its position at the opening of the hearing. He will have two minutes to do this.
6. After each group has stated its view, all the groups for and against the electric company's plan will take turns arguing their cases. There will be enough turns to allow all members of your group to speak.

If the hearing and class end at the same time, the Board members may be able to get together and reach a decision sometime before class the next day.

THE ROLES

Officials of the State Health Board

You will hold a public hearing on the Wattson Electric Company's power plant. You will decide if the company can release heated water into Serena Lake.

It is your responsibility to protect the rights of everyone when water is concerned. Like most citizens, you are concerned about the quality of your environment. You feel that everyone has the responsibility to keep the lake as clean and unpolluted as possible. Yet, you feel that industries have rights, too. They should not be so controlled that they find it difficult to operate.

The problem of thermal pollution is new and complicated. It will take all your skill and wisdom to reach a fair decision. Therefore, you need to study the data on pages 313–323 so that you can understand the position of each group. The data will also help you ask good questions.

One of you will act as chairman for the hearing. The witnesses will try to convince you to agree with them. But you will permit them to talk to each other, too. However, you will insist that only one group may speak at a time.

The price of progress?

How the chairman should conduct the hearing

a. Call the hearing to order.
b. Introduce yourself and the other Health Board officials. State the reason for holding the hearing.
c. Ask one player from each group to introduce his group. Ask him also to state briefly his basic position on the issue. Allow just two minutes for each group. Call on the groups in the following order:
   1. Wattson Electric Company
   2. Citizens Committee to Save Serena Lake
   3. State Industrial Association
   4. Friends of the Environment
   5. Majority of city council
   6. Minority of city council
d. Next, call on each group in the same order to support its position. Allow each group three minutes to present its data. Health Board members may interrupt to ask questions at any time. Let the groups for and against the permit take turns challenging each other's positions and defending their own. They can have up to four turns each.
e. After the debate, give each group a chance to make a final statement. Allow each group two minutes.
f. Meet privately with the Board members to reach a decision. You can take up to 15 minutes. Then report your decision to the groups. Explain why you reached that decision.
g. Adjourn the hearing.

Officials of the Wattson Electric Company

You are convinced that atomic power is the key to a vast new supply of cheap electricity. Therefore, you are convinced you should build the atomic power station on Serena

Lake. It is described in the Data Bank. You consider your company's main goal to be to provide the best service for the greatest number of people. You want to provide enough electricity for all your customers at the lowest possible price. You also feel responsible to the company's stockholders. The atomic power station will add to the company's profits.

Local people are using more electricity than your stations now produce. You have to buy electricity from other power companies for your customers. Every year they have more appliances, such as air conditioners, electric stoves, toothbrushes, scissors, and can openers. Many new homes are being heated with electricity. The demand for electricity keeps growing. Figure 21.13 shows the federal government's predictions for future power needs.

A study of the lake was made for two years. You think it shows that Serena Lake is a good place for the station. For example, there is a large enough supply of cold water to cool your steam-turbines. Opponents of the power station claim it will harm the environment of Serena Lake. You

FIGURE 21.13
This graph shows the electrical power needs of the United States. The predictions were made by the Federal Power Commission. The average family of five now uses 5,000 kilowatt hours a year.

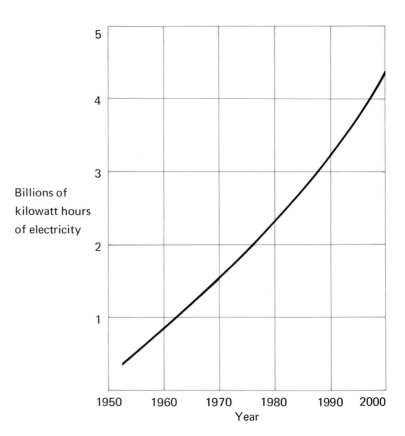

disagree. You intend to protect the environment, as well as anyone might. After all, the success of your company partly depends on its being a good citizen.

You agree that heat released from the station will cause *some* changes in the lake. But, the increase in the average summer temperature of the whole lake will only be about *one degree*. This slight increase in temperature will be cancelled by the following winter. Therefore, the lake will not be harmed.

Your company is conducting more research to learn about possible harmful effects to the lake. If the research clearly shows such effects, you would change the design of the station to avoid doing harm. However, data is now being gathered on the effects of the power station already operating on Serena Lake. You are convinced that this data supports your plans for the atomic station.

You don't want to build a cooling pond or cooling towers for your station. The land around the lake is too hilly to build a cooling pond, and the towers are big and ugly. There is no data on the effects of large cooling towers in areas with long, cold winters. They will surely create fogs in the area and bring public criticism. In fact, if you are forced to use artificial cooling systems, you may decide to build the station in a nearby state.

You are very concerned about a delay in getting the station built. A delay will increase costs by as much as 30 million dollars within a year.

State Industrial Association

The Association represents many of the state's industries. The Central City area has steel, oil, chemical, auto, and many smaller industries.

Your organization strongly supports Wattson Electric's plans for the power station. It will provide cheap electricity.

Industry around Central City is growing. Industry's need for electricity is increasing, too. When industries expand, they create more jobs and prosperity.

There are many growing companies that now wish to move into the area. Several of these are the Midwest Aluminum Company, Comfort Air Conditioners, and the General Glass Company. But they are waiting to see whether the new power station will be built. They know their profits will rise if electricity prices fall.

The Citizens' Committee demands a cooling pond or cooling towers for the new power plant. The Committee doesn't want Serena Lake to be heated up. You agree with the power company that this demand is too strict. It might hurt the economy of the area: Cooling methods cost extra money. If industries pay more for electricity, customers will have to pay more for goods. No new industries will move to Central City and then there may not be enough new jobs. You say the benefit to the lake environment won't be worth this cost.

Finally, you think it's up to the opposition to prove that the environment will be harmed. Otherwise, industry should not be told what to do or how to do it.

The majority of city council members

You are residents of Central City. You have lived and worked here all your lives. Recently, most of your fellow council members voted to support the present plans for the power station.

You are excited about the money that will come in from the construction and operation of the station. It will also create many new jobs. The area's economy will get a big boost. Local builders, building suppliers and manufacturers, realtors, and landowners will profit handsomely.

The Citizen's Committee is trying to force the company to use a cooling pond or cooling towers. You fear that this will increase the cost of electricity. Businesses might be forced to leave the area. Besides, the pollution of Serena Lake has been going on for years. It is unfair to pick on the power station. You accuse the Citizens' Committee of blaming Wattson Electric for everyone else's sins.

None of you want the purity of the lake water to be affected by the station. You are sure that the company will try to keep the quality of the water high. You want to give the station a fair chance. You think that any hasty decisions to change it would hurt the community. One of your frequent comments to local citizens is, "For once in our history, let's try to get something that *pays* local taxes and doesn't just *spend* tax money."

Citizens' Committee to Save Serena Lake

The Citizens' Committee was formed about a year ago. It has more than 200 members. Sportsmen and conservation groups also plan to join.

You know that Central City needs a new power station, but you are opposed to the atomic station. You think it will give off too much heat to the lake. You are sure the heat will speed up the aging process of the lake. You have data showing the harmful effects of heating an aquatic environment. You argue that it might not be possible to correct the damage once it occurred. You want *no* heat added to Serena Lake unless the Wattson Electric Company can prove that the heat will not harm the aquatic community.

Local scientists have advised you that the station should not pipe hot waste water directly into the lake. Therefore, you insist that the company use other methods, such as a cooling pond or cooling towers. The company opposes your position. You think it is because the company doesn't want to recognize the danger. You are sure that the whole lake will be harmed. Also you think that Wattson Electric wants to keep profits high and not spend the additional money.

You know that the company itself is having research conducted to obtain more data about the effects of heat on Serena Lake. They are gathering data at the power station now operating on the lake. However, you don't think this is a good procedure. You think data should be collected from several places around the lake. This will show what thermal pollution could do to the whole lake.

Since Central City depends on the lake water for drinking, you have another concern. You predict that changes in the lake will make the water unfit for drinking.

Finally, you urge the company to stop constructing the station until it can guarantee that the lake will not be harmed. You are sure a cooling pond or cooling towers will provide the guarantee. It makes you angry to think that the state government spends millions of dollars to clean up water pollution, and then will permit a power station to make more pollution. You insist that individual citizens have the right to prevent pollution.

Friends of the Environment (FOE)

Do people really need electric scissors, knives, and shoe-polishers? You are sick of people being told they need things that they can easily do without. (See Figure 21.14.)

The price of progress?

America is TURNING ON to Electricity

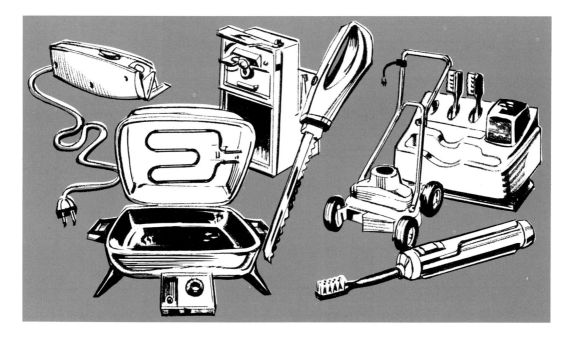

Think of how these electric appliances can make your life cleaner! easier! more enjoyable! healthier! Your electric company supplies the power! All you need are the appliances! Then you're on your way to electric living!

ELECTRICITY MEANS PROGRESS

FIGURE 21.14
Here is an advertisement put out by the Wattson Power Company. Do you agree with FOE that these appliances are unnecessary?

The price of progress?

FOE is a group with members all over the state. Your members feel it is your duty to protect the environment. You want to pass on the environment to your children in as good or better shape than you found it.

Your organization is strongly opposed to building the atomic power station on Serena Lake. In fact, you are against having any station located on the lake. This includes the power station already operating there. You believe that man cannot go on adding heat to his environment without damaging it. If he continues, you say, he will destroy himself. The data that shows the effect of heat on aquatic communities makes you believe this.

You are certain that the solution to the thermal pollution problem is for people to use *less* electricity. Then new power stations would not be needed. You accuse Wattson Electric of deliberately creating the need for more electricity to make higher profits. As evidence, you point to the company's advertisements to increase the sales of electrical appliances. You want to see more research done to make electricity from the wind, or ocean tides, or sunlight.

You accuse Wattson Electric and anyone who supports the station of being irresponsible. You say they only care about using the lake to make money. But you feel that the lake should continue to exist, unchanged. Everything in nature doesn't exist for man to profit by.

Minority of city council members

You are residents of Central City and the younger members of the City council. All of you have lived and worked here less than ten years.

The price of progress?

You voted against sending a statement to the Health Board in support of the power station. But you were outnumbered: Everyone else on the city council voted in favor of the station. You want the station to be built here, but not as it is now planned. You consider it your duty to protect the lake. You have read scientific reports describing what happens to lakes when the water is warmed. Frankly, this data frightens you.

You accuse Wattson Electric of having its own research conducted to prove that heat pollution will not harm the lake. Otherwise, why would they begin to build the plant before all the data is gathered? Also, if they really wanted impartial research, they would have data collected from more than one place in the lake.

You realize that if the station is built here, it will pay badly needed taxes. But think about how much it could cost to harm the lake! The pleasure local people get from swimming, fishing, and boating can't be measured in dollars and cents. The tourist business would die. That might have as bad an effect on the economy as the loss of the station. You think a cooling pond or cooling towers is the answer. The additional cost would be worth it.

You claim that there is a lot of opposition in Central City to the station. Therefore, you support public hearings like this one. They give the public a chance to fully understand the issue. You insist that Wattson Electric must explain its operation to the public.

The data bank

Entry 1—Serena Lake, now

PHYSICAL CHARACTERISTICS

Serena Lake is the smallest lake for which an atomic power station has ever been planned. The water in the lake hardly moves. It takes about ten years for a drop of water to travel from one end to the other. Water finally flows slowly out of Serena Lake through a large stream.

AGING

From May to November, algae grow in the warm water near the surface. The water temperature in the summer months

314   The price of progress?

FIGURE 21.15
Serena Lake is on the edge of an industrial area, but much of the shoreline is still wooded.

FIGURE 21.16
This cross-section of Serena Lake shows how the surface and the bottom temperatures differ.

is about 70°F (21°C). Light is available near the surface, along with necessary minerals and gases (oxygen and carbon dioxide). Like most lakes, Serena Lake is gradually filling up.

This is what happens to lakes as they get older. Nitrogen and phosphorus, minerals important for plant growth, are washed into the lake by water running off the land. These added minerals make algae and other plants grow faster and faster. Then the animals in the lake have lots of food. Their populations grow also. As all these organisms die and sink to the bottom, the lake slowly fills up.

The algae also make the water greener, so more sunlight is absorbed. This warms up the water so the algae grow even faster. These changes are part of the natural filling up and aging of the lake. This process may take thousands of years.

The area around Serena Lake is being developed for housing, industry, and agriculture. As a result, more soil washes into the lake. Pollution is dumped into the lake, too. Algae can grow on the new materials. In fact the algae is growing so fast that it is making the water murky. Recent studies show that Serena Lake is aging unnaturally fast. Twenty years ago, the water was clear down to 15 or 20 feet. Today, it is clear to less than 10 feet. This makes swimming and water skiing more dangerous. Many of the people who live near the lake are worried. Their incomes depend on tourists and vacationers who come to the lake to go swimming, boating, and fishing.

HOW THE LAKE IS USED

People use Serena Lake for recreation, city and industrial water needs, and city sewage disposal. The estimated values of these uses are from 6 to 11 million dollars a year.

THE PRESENT POWER PLANT

Wattson Electric already has one small power station on the lake shore. It uses coal to generate electricity. Waste heat from this station goes directly into the lake. Wattson Electric Company did some studies to see what effects waste from the station has on the lake. Researchers found that the waste heat did not seem to affect the lake.

The water for this station is taken from a depth of 45 feet. Water for the atomic station will come from 100 feet down. The new station will use at least four times more cooling water than the existing station.

316 The price of progress?

Entry 2—The proposed power station

OPERATION

The Wattson Atomic Power Station will be one of the largest power stations in the country. It will also be the first atomic power station built on the shore of a lake. Company officials estimate it will take five years to build and cost 125 million dollars. Figure 21.17 shows how the station will operate.

This atomic power station will give off about 50 per cent more heat than a station which uses coal, oil, or gas for fuel. But the station will not give off any smoke, dust, or bad smells. It will produce another kind of pollution, though. About once a month the cooling water pipes have to be

FIGURE 21.17
a. *How an atomic power plant works. The nuclear reactor produces heat. Water is boiled by the heat to make steam. The steam spins the turbine generator to make electricity. Afterwards the steam is cooled back to water. Cool lake water removes heat from the steam and is returned to the lake as warm water. Lake water doesn't touch the reactor.*
b. *The water for cooling the steam will be taken from 100*

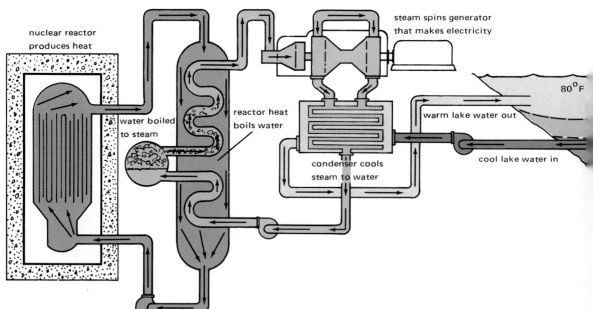

a

The price of progress?

cleaned out with a strong chemical. This chemical is flushed into the lake.

ECONOMICS

To construct the station will take about 1,000 workers. These workers will earn at least 20 million dollars. Materials to build the plant will cost 15 to 20 million dollars. These supplies will be purchased locally. This means that the area will receive about 40 million dollars worth of business.

After the station is built, the power company will pay taxes to the city. These taxes will be spent to build roads and schools and make other improvements. Finally, about 60 skilled workers will be needed to operate the station.

feet down in the lake, where the water is as cold as the inside of a refrigerator. When this water comes out of the power plant, it is 25°F hotter. The heat in the water spreads out in the lake and eventually goes into the air. The power company says that the overall temperature of the lake will be raised only one degree. (Is the overall rise in temperature the only figure that matters?)

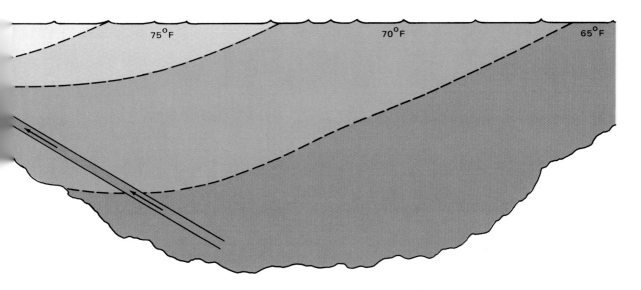

b

The price of progress?

Entry 3—Ways to remove heat from power plants

DIRECT FLOW

The cheapest way to get rid of waste heat from the power station is to pump lake water through the cooling system. This method is called the direct flow method. It is shown in Figure 21.17b. All the heat goes into the lake.

COOLING PONDS

A cooling pond like the one in Figure 21.18 is one way to avoid adding heat to Serena Lake. The pond would be built near the power station. It would be filled with lake water.

FIGURE 21.18
This cooling pond could be built for the atomic power plant, to hold warm water until it is cooled. Then the cool water could be reused.

The deep end of the pond (A) would contain cool water. This water would be pumped to the station to cool the steam. The heated water would be let out into the shallow end (B). It would cool off before being used again. Its heat would slowly go into the air. Notice that the Figure shows some other ways this pond could be used.

A pond for the Wattson Power Station would have to cover about two square miles. It would cost about 2.5 million dollars to buy the land, dig it out, and flood it. The pond would not be expensive to run, and the water of Serena Lake would not be heated. However, this method may not be practical for Serena Lake. The land at the power station site is hilly.

COOLING TOWERS

Another way to get rid of waste heat is to use cooling towers like those in Figure 21.19 on the next page.

In a "wet" cooling tower, hot water from the station is sprayed over metal plates. Cool air from outside blows up past the plates. Some of the hot water is evaporated by the cool wind. Evaporation cools the remaining water. The cool water can be returned to the station to be used again. Or it can be piped into the lake. The evaporated water may form a cloud over the area. In cold winters, it can create heavy fogs and icing.

A "dry" cooling tower works like a car radiator. Hot water passes through small metal pipes. A fan blows cool air past the pipes. This wind cools the water inside. There is no evaporation and less danger of fog. The cool water can be returned to the station for reuse, or emptied into the lake.

The two "wet" cooling towers in Figure 21.19 are each over 400 feet tall. They cool thousands of gallons of water a minute. Hot water enters the tower at 80°F. It can be cooled to 50°F by the time it leaves.

Cooling towers would not heat up the lake. Officials of the electric company say that these towers are very expensive. No one disagrees with them. "Wet" towers would cost about 10 million dollars to build. "Dry" towers would cost about 25 million dollars. These costs would increase the price of electricity. The average home owner's electric bill would be at least 50 cents more a month. Industrial users of electricity would pay much more than that, since they use more electricity.

FIGURE 21.19
"Wet" and "dry" cooling towers.
a. "Wet" cooling tower.
b. "Dry" cooling tower.
c. Wet towers being used.

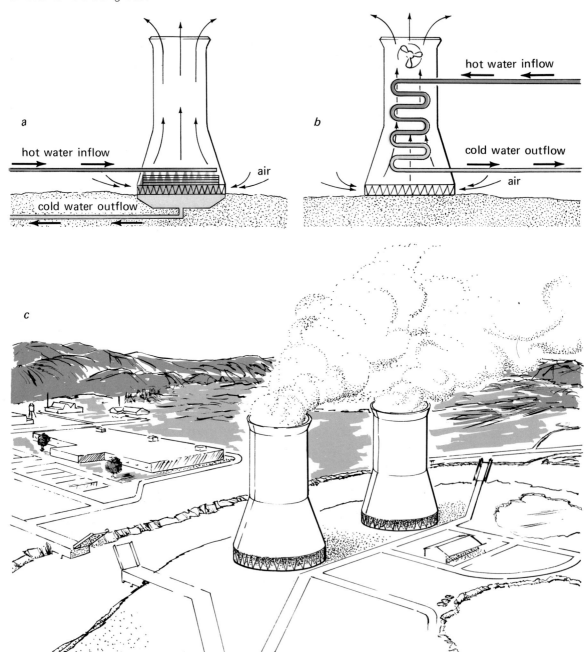

Entry 4—The effect of adding heat to an aquatic community

RATE OF LIFE PROCESSES

Increasing water temperature makes aquatic plants and animals grow faster. It speeds up the use of food, and the rate of gas exchange, and heartbeat in animals. The organisms grow faster, but they do not grow as large or live as long as in cooler water. When temperatures get too high plants and animals begin to die off. See Figure 21.20.

EFFECTS ON ALGAE

The normal growing season for algae and other plants in Serena Lake is from May to November. Warming the lake water will lengthen the growing season. This will speed up the aging process of the lake. However, the exact speed of

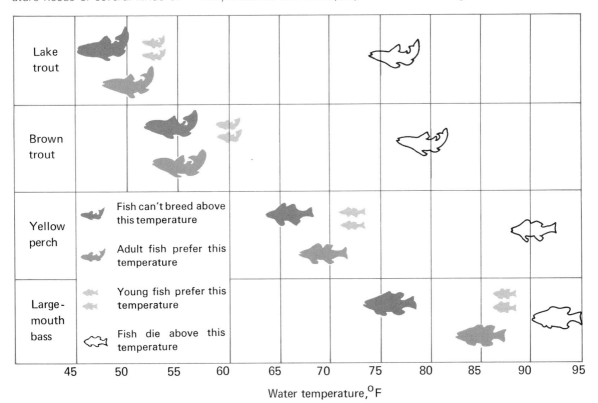

FIGURE 21.20
The diagram shows the temperature needs of several kinds of fish living in Serena Lake. Carp and catfish grow well at higher temperatures, but most people don't enjoy catching or eating them. What will happen if Serena Lake gets warmer?

FIGURE 21.21
*These kinds of algae are often found living together in cool lake water. Above 68°F, the kinds of **diatoms** (DY-uh-toms) decrease and the numbers of blue-green algae increase. Most kinds of green algae need moderately warm water, about 86°F. Some biologists think that blue-green algae are not a good source of food for algae-eating organisms. These organisms seem to grow better when they eat diatoms. Also, many blue-green algae produce an unpleasant smell and taste in water. Some are poisonous to clams. What will happen to the algae populations if Serena Lake gets warmer?*

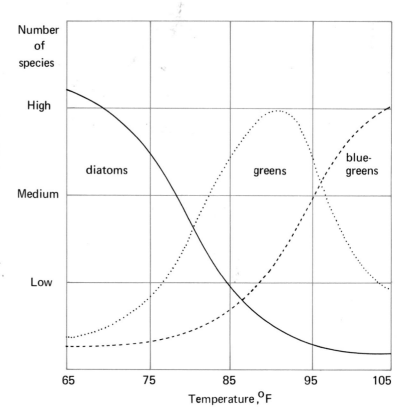

aging cannot be predicted. The graph in Figure 21.21 shows how water temperature affects algae populations.

EFFECT ON OXYGEN

The hotter water is, the less oxygen it can hold. At higher temperatures, bacteria and fungi will decay dead organisms faster. Decay uses up oxygen. Then there is even less oxygen in the water. Most desirable aquatic organisms need a lot of oxygen. Some pest organisms grow well with very little.

EFFECTS ON FOOD WEBS

The best range of temperature is one that keeps the whole aquatic system healthy. Game fish such as trout usually need the cooler water near the bottom of the lake or river. Minnows and young fish are found in the warm shallows. Fish, frogs, clams, crayfish, plants—each organism has certain temperature needs. Figures 21.22 and 21.23 show healthy and unhealthy aquatic environments.

The price of progress?

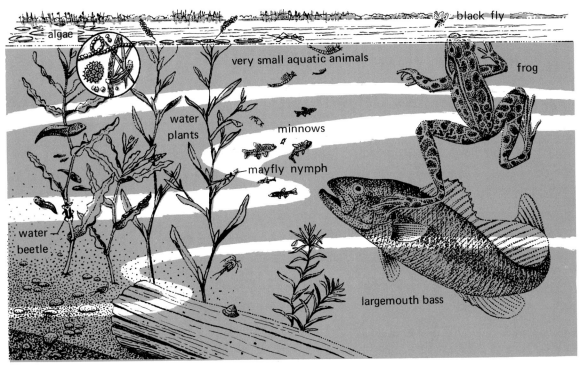

FIGURE 21.22
Many kinds of organisms live in the healthy community shown here. Warming the water can kill off some kinds of young insects, shrimplike animals, and other plant-eaters. Many fish and frogs that normally eat them starve. Then there are fewer game fish.

Post-game discussion

After you have finished the hearing, discuss it with your classmates. The most important data in the game is not in the book. It is data on how people behave when a decision has to be made. Analyze your own actions in the debate. Your teacher will help get you started. Afterward, discuss these questions.

a. Did players support their positions by logic and facts, or were they emotional? Does emotion belong in discussions of science and business?
b. In your own opinion, what was the most convincing argument made to the Health Board?
c. Should special interest groups such as a Citizen's Committee, Industrial Association, or power company be allowed to influence the Health Board? What should be the role of special interest groups?

The price of progress?

FIGURE 21.23
An unhealthy environment is shown here. Some organisms in it make conditions that kill off even more plants and animals.

    d. Were there any winners? How do you decide?
    e. What problems arise in a conflict between desires for prosperity and an unspoiled environment?
    f. If you held a hearing with a new Health Board, could you change the outcome? How?

Mastery Item 21-1

How will you invest your money now?

Turn to the thermal pollution gameboard again. Then check your records for the amount of money you had at the end of Problem 21-1. Now you will play the final round of the game (Now–20 years in the future).

    Your task is to decide what to do with your money. You can use ideas you have from both Problems. The game rules do not apply to this Period. You are free to invest as much or as little of your money as you wish. If you like, you can invest in businesses not listed on the gameboard.

On a separate sheet of paper list the following information:

a. The amount of money you invest and the names of the businesses.
b. What you did with your money if you did not invest any.
c. The maximum number of degrees you would be willing to increase the water temperature to double and triple your money.
d. How you think your investment will affect thermal pollution.
e. The ways, if any, that you might get rid of waste heat without increasing thermal pollution.

Key

There are no "right" or "wrong" answers to this item. Discuss and defend your decisions with the class.

FIGURE 22.1
This is a limb from an elm tree. The bark has been removed to show where beetles lived under the bark when the tree was alive. The beetles "carried" the Dutch elm disease that killed the tree.

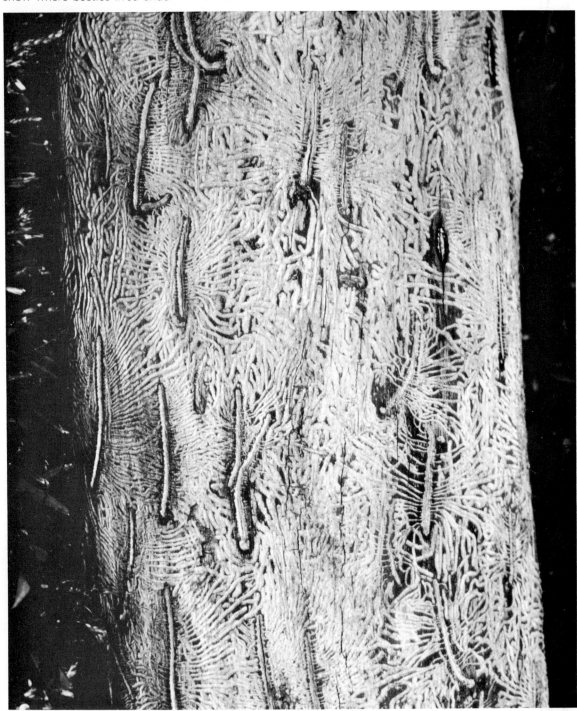

Investigation

22 What can we do about pests?

What do we do to our environment when we try to control pests? Do we know all the effects? Is there a hidden cost for having perfect fruits, beautiful lawns, or weedless roadsides? Is it really possible to have a pest-free, shiny, controllable environment?

What is a pest anyway? Pests are living things that can cause harm or discomfort. The harm can be real or even imagined. We call weeds pests just because they grow where we do not want them to grow. A mosquito is a pest when it feeds on us and causes us to itch and scratch. Some kinds of mosquitoes also carry diseases like **malaria** (muh-LAIR-ee-uh) and yellow fever.

Flies and mosquitoes can be considered "natural" pests. That is, they bother us, even in a natural environment. On the other hand, dandelions are often called pests because they grow in a part of the environment someone wants to change into a lawn. In a natural environment dandelions are just wildflowers. When we decide that the original inhabitants of a place, either plants or animals, are undesirable, we "create" pests.

How many of our pests have we ourselves created by our activities? Make a list of all the "natural" and "created" pests you can think of.

In this Investigation you will study how man has tried to control different pests. You will find out about some different pest control programs. You will see if they have done what man predicted they would do.

Problem 22-1

Controlling pests

Ever since man farmed large areas of the earth's surface, his chief competitors for the crops have been insects. His crops are attacked by hordes of "created" insect pests. In a field or in the woods an insect has to compete with many different organisms for food and space. It has natural enemies, too. Therefore, its numbers are kept in check. But farming destroys many organisms and their homes and food supplies. Often, the few kinds of insects that remain develop large populations. They feed on the planted crops. Man then goes to war against the enemy he "created."

FIGURE 22.2
Giant spraying programs are used to try to control pests. This plane is spraying an African city to kill malaria-spreading mosquitoes.

What can we do about pests?

Four reports describing a variety of different pest problems follow. Share the study of these reports with your classmates. Your teacher will help you decide how to divide up the work. You will then try to choose some of the best ways to solve pest problems.

Report 1. unexpected results

The weapons used most often in the war against insects are chemical poisons. They are called **pesticides** (PES-tuh-sides). Man has covered thousands of square miles with pesticides to try to protect his crops.

A pesticide is sprayed over an area to control what is called "the target insect." These pesticides usually do decrease the population of the target insect. But what do they do to the rest of the environment? It is often impossible to predict *all* the effects of spraying. Here are some recent examples of unexpected results from spraying.

FIGURE 22.3
"This is the dog that bit the cat that killed the rat that ate the malt that came from the grain that Jack sprayed."

What can we do about pests?

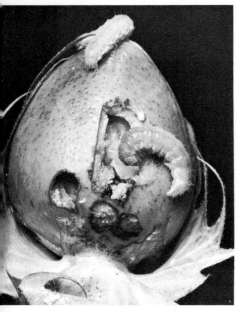

FIGURE 22.4
Bollworms destroying a cotton boll. This boll won't produce any cotton fibers.

FIGURE 22.5
Fire ants feeding on a flower bud

THE BOLLWORM IN NORTH AFRICA

In North Africa cotton fields were sprayed with DDT to control several kinds of insects which feed on the leaves of cotton plants. The early treatment was successful. So, the growers increased the amounts of pesticide sprayed on their cotton fields. Then the unexpected happened. Another, more destructive insect pest in the cotton fields, the bollworm, multiplied. Apparently the spray was killing the enemies of the bollworm. The unsprayed cotton was suffering less damage than sprayed plants. True, some of the leaf-eating insects had been eliminated. But the bollworms destroyed more cotton than leaf-eating insects ever had.

THE FIRE ANT IN LOUISIANA

Another example of unexpected results came from an attempt to control the fire ant in Louisiana. The fire ant was a nuisance. But it was not an important crop pest. However, the state decided to kill off the fire ants with the pesticide, **heptachlor** (HEP-tuh-klor). The heptachlor spray also killed off insect enemies of the sugar cane borer. The sugar cane borer, one of the worst enemies of the sugar cane crop, multiplied out of control. The damage to sugar cane crops in Louisiana was so great that farmers sued the state for careless use of poison sprays.

THE GOLDENROD ON ROADSIDES

Similar experiences have resulted from attempts to control plant pests. Poisons were sprayed on plants, such as goldenrod along roadsides. They killed desirable plants, also. Then new plants, which were even less desirable than goldenrod, moved into the vacant spots.

THE SPRUCE BUDWORM IN OREGON

Here is a last example. Our western national forests are an important source of lumber. Several kinds of evergreen trees in these forests are preyed on by the spruce budworm. The young spruce budworm tunnels into needles and buds on the ends of branches. If there are lots of insects in a tree, they can destroy all its needles. Nearly three million acres of Douglas fir have been seriously damaged in southeastern Oregon alone since 1945. Other evergreens in the forests of northern Maine have also been badly damaged.

FIGURE 22.6
Spruce budworm larvae feeding on a Douglas fir cone

FIGURE 22.7
These tips of fir branches have lost their needles and died because of budworm attacks.

FIGURE 22.8
Red spider mites can be a serious forest pest.

To control the spruce budworm one year, the United States Forest Service planes sprayed 885,000 acres of forest lands with DDT. The program looked like a success. Most of the budworms were destroyed. The next summer, very few budworms were found, but the trees were dropping their needles. Thousands of acres of forest were turning brown. Damage was as bad or worse than the damage by budworms. What happened? A close look revealed the following information.

1. The red spider mite, which also preys on evergreen trees, was not killed by the DDT.
2. Animals such as the ladybug beetle, which eats spider mites, were destroyed.

3. Spraying irritated the spider mite colonies. The irritated but unkilled mites scattered out to places where they would not be disturbed.
4. Finally, the spider mites, with no enemies to control them, were doing great damage to the trees.

What went wrong in the three cases you just read about? Does the idea of food webs help you understand? What do you think should be done before a large spraying program is started?

Report 2. poisons turn up everywhere

Up until this century man has not been able to fight deadly diseases like plague, malaria, or yellow fever. Malaria is a disease spread to man by a kind of mosquito. As recently as 1920 there were 200,000 cases of malaria each year in the United States. There were over 60,000 cases a year until 1945. Then the mosquitos that carry malaria were attacked with the insecticide, DDT. The number of new malaria cases dropped rapidly. By 1950 there were only 2,000 cases per year.

DDT has been helpful in reducing other diseases carried by insects, too. Altogether it has been used in the control of 38 serious diseases. Some people estimate it has saved 25 million lives.

The use of DDT on farmlands has helped increase food production. During the 1960's, man used DDT to control over 100 kinds of insects that feed on crops and livestock. DDT helped man control pests like army worms, blister beetles, caterpillars, corn borers, and many other damaging insects. As a result man eats fewer insect-damaged foods.

If DDT has done such good work, why has its use been forbidden in many places, including the United States? Because it is now known that DDT and similar pesticides poison many other kinds of animals, including man. These chemicals are especially dangerous because they are spread through the environment by wind or running water. They stay actively poisonous for many years. Such chemicals, like DDT, are often called "hard pesticides." It is estimated that over one billion pounds of these chemicals are now in the earth's water, air, and soil.

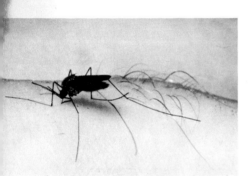

FIGURE 22.9
The Anopheles mosquito carries malaria. The malaria microbes will enter this person's bloodstream.

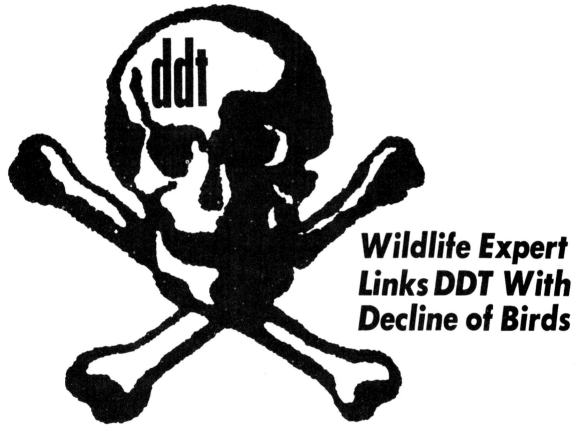

FIGURE 22.10
"Hard pesticides" are now found everywhere. Man is beginning to find out how harmful they are to all living things.

334   *What can we do about pests?*

Fourth-order consumers

Third-order consumer

First-order and second-order consumers

Producers

FIGURE 22.11
This ocean food web shows how seals have been poisoned.

How do pesticides travel from your backyard to the middle of the ocean?

POISONS IN ATLANTIC SEALS

How can hard pesticides poison animals that haven't been sprayed with them? How can hard pesticides spread across the world?

Seals live in the ocean where no hard pesticides, like DDT or **dieldrin** (DEEL-drin) have ever been sprayed. Figure 22.11 shows part of the food web for seals. As you can see, seals are fish eaters. These fish, in turn, eat smaller organisms in the ocean. Seals that live in the ocean between Scotland and Canada were examined in a recent study. High concentrations of DDT and dieldrin were found in their flesh. Where do you think these poisons came from? Recall how water gets into oceans.

FIGURE 22.12
A baby eagle sits beside a damaged egg in a nest. Soon there may be no more bald eagles. DDT keeps them from producing young.

POISON IN CONNECTICUT OSPREYS

A story closer to home concerns the fish hawks, or **ospreys** (OSS-prayz), on the Connecticut River. In 1954, there were about 150 pairs of these beautiful birds living and breeding in the river valley. In 1964, there were less than 15 pairs. Few of these birds now produce young. Biologists believe that DDT can prevent female birds from laying eggs. They know it causes some birds to lay eggs with thin shells. Then the eggs don't hatch. In the ospreys both the unhatched eggs and adult birds contained a large amount of pesticides.

Where does the DDT come from? It is not sprayed on the birds. It comes from the food they eat. Figure 22.13 shows how the amount of poison increases in a food chain. A plant takes in the chemicals from its environment. When an animal eats many such plants, he collects the DDT from all of them. Most of it is stored in the animal's fat. An animal at the end of the food chain, the osprey or seal for example, can slowly collect a lot of poison. This will upset its normal life.

FIGURE 22.13
How do ospreys get poisoned? The triangles show the amount of DDT in each link of their food chain.

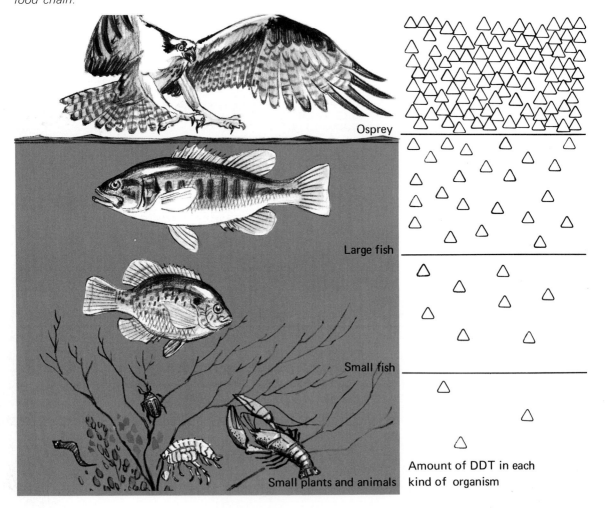

What can we do about pests?

POISON IN THE CLEAR LAKE GREBES

Clear Lake, in Northern California, is a popular vacation spot for people living near San Francisco. At one time, there were about 1,000 pairs of **grebes** living around Clear Lake. Grebes are fish-eating ducks. They build floating nests on the edges of lakes.

In the winter of 1954, the grebes of Clear Lake began to die. More than 100 were found dead. By 1969, only about 30 pairs were left. They made nests, but no young grebes have been observed since that time.

An investigation after the death of the grebes showed that resort owners and tourists at Clear Lake had been bothered

FIGURE 22.14
A grebe standing on its floating nest.

What can we do about pests?

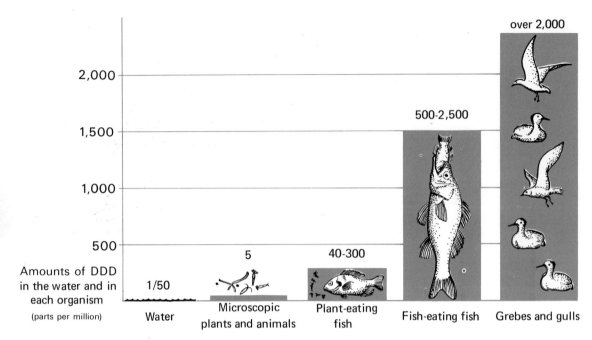

FIGURE 22.15
This graph shows the amount of DDD found in organisms of Clear Lake. Why did the grebes contain so much pesticide?

by swarms of **gnats** (NATZ). These are tiny biting insects which breed in the lake. People wanted to control the gnat population to make the lake more pleasant for themselves. It was decided to spray the lake with a new chemical, DDD.

But before spraying, the lake was carefully surveyed. The total amount of water in it was calculated. There was some evidence that one part of DDD to every 70 million parts of lake water would kill the gnats, but not the fish. The lake was first sprayed with this amount of DDD in 1949.

Control of the gnats at first was good. In a few years, however, swarms of gnats returned again. Clear Lake was sprayed several more times over the next eight years. The amount of DDD was increased to one part of DDD per 50 million parts of lake water. But this amount of DDD was supposedly low enough not to harm the fish.

Did the DDD have anything to do with the death of the grebes? Could a "safe" level of DDD kill them? Figure 22.15 shows how much DDD was found in the plants and animals of Clear Lake. The food chain in Clear Lake was now a

What can we do about pests? 339

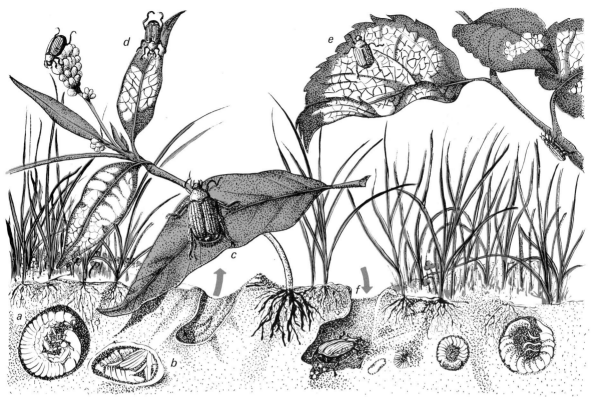

FIGURE 22.16
Japanese beetles are pests both as young grubs and as adult beetles: a. *grub in the spring, eating roots;* b. *the grub resting as it changes into a beetle;* c. *an adult beetle coming out of the ground;* d. *beetles feeding on smartweed;* e. *beetles feeding on apple leaves;* f. *a female beetle laying eggs which will hatch into grubs.*

poison chain! The poison had moved from plankton to fish to grebes. The lake had been sprayed with only one part of DDD to 50 million parts of lake water. Do you think there is an absolutely safe amount of hard pesticide to add to any environment?

POISON IN WORMS, BIRDS, SQUIRRELS, AND CATS

Japanese beetles were first seen in the United States in 1916 in New Jersey. They probably entered the United States on plants brought from other countries.

Japanese beetles damage orchards, crops, and lawns. The young beetles, called grubs, live underground. They eat roots. The adult beetles eat the leaves of many kinds of plants and trees. The Department of Agriculture made an estimate that Japanese beetles cause 10 million dollars damage in the United States each year. Two different methods to control the Japanese beetle have been tried.

One attempt to control Japanese beetles was started near Sheldon, Illinois, in 1954. The United States Department of Agriculture and the Illinois Agriculture Department used dieldrin, a hard pesticide. By 1961, 131,000 acres had been sprayed.

Dieldrin soaked into the soil and killed many beetle grubs and other insects. Earthworms also absorbed large amounts of poison and many died.

By 1961, the entire bird population, including starlings, brown thrashers, meadowlarks, grackles, and pheasants, was practically wiped out around Sheldon. Robins became rare. Only a few of the birds that survived produced eggs. No nests containing young birds were found during the next summer.

In addition, ground squirrels were almost killed off. The bodies were found all twisted up, as though they had died a violent death by poisoning. Dead rabbits and muskrats were also found in the treated areas. The town of Sheldon lost its whole fox squirrel population. Ninety per cent of all the farm cats died during the first season of spraying. Why do you think that all these different kinds of animals died?

Although everything else was dying, the Japanese beetle still continued to multiply and do damage. And it continued to spread farther west.

In another attempt to control the Japanese beetle, some Atlantic states decided not to use a chemical poison. They spread a kind of bacteria that causes disease in the beetles.

This kind of bacteria does not attack other kinds of insects, earthworms, mammals, or plants. When they are eaten by a Japanese beetle grub, the bacteria grow rapidly in its blood. They make the blood turn white. The disease has been called "milky spore disease."

Between 1939 and 1953, the bacteria were added to the soil of 14 states. By 1945, milky spore disease was found in the beetle populations of Maryland, Delaware, New Jersey, New York, and Connecticut. In the eastern United States, where this program began, the milky spore disease bacteria stays in the soil. Year after year it seems to be keeping the beetle populations well under control.

FIGURE 22.17
A Japanese beetle larva is normally dark-colored (left). A larva infected with milky spore disease (right) is much lighter in color. Millions of white spores multiply in its blood and soon kill it.

a   b

The government agencies responsible for the spraying program in Sheldon, Illinois, claimed that spreading milky spore disease was more expensive than spraying and did not show immediate results. What is your opinion? Which treatment would you prefer in your state?

POISON IN MAN

Because it does such a good job, DDT is still used in many places around the world. Does it get into food chains that end with humans? Data from a recent survey is shown in Figure 22.18. Do you think your body contains DDT? Even people isolated from modern civilization contain some DDT.

Do you use hard pesticides in your everyday life? Look at the cans of insect and weed killers in your home or in stores. What chemicals do they contain? Are DDD, DDT, DDE, aldrin, endrin, methachlor, lindane, chlordane, dieldrin, heptachlor, or epoxide listed on the label? If so, then you, or your family, or your community are contributing to the amount of long-lasting, poisonous chemicals in the environment. And you may be contributing to the pesticides in your own body.

Can these chemicals harm man? Many of them are known to be very poisonous in large amounts. Nearly all of the foods we eat contain very small amounts of pesticides. No one knows for sure if damage can result from taking in small amounts for a long time. Should we allow these poisons to increase in our bodies while we wait to find out?

Should we try to stop the spread of poisonous chemicals in the environment? When is the use of insecticides a good idea? When should it be stopped? Who should make these decisions?

FIGURE 22.18

This table shows the amounts of DDT in people of different countries. When you eat food containing DDT, it stays in the fat in your body for many years. Why do you think that people from India have the most DDT in their bodies?

Country	Amount of DDT in the body (*parts per million*)
U.S. (average)	11.0
Alaska (Eskimo)	2.8
England	2.2
West Germany	2.3
France	5.2
Canada	5.3
Hungary	12.4
Israel	19.2
India	12.8–31.0

Source: George, W. "Toxic Substances and Ecological Cycles," *Scientific American*, March 1967, Vol. 216, p. 31.

Report 3.
creating "super-pests"

We have been able to breed horses to run faster and cows to give more milk. Our chickens lay more eggs than 50 years ago, and our apple trees produce larger crops of redder apples. These improvements in organisms are produced by **selective breeding.** In selective breeding only the organisms with the most desirable characteristics are mated. Selective breeding over many generations can produce organisms that are very different from their ancestors.

Without meaning to, we have also been breeding superior pests. Here are some examples.

"SUPER-PESTS" CARRYING DISEASES

In 1943 in Italy, a campaign was started to control diseases that were spread by insects. People were dusted with DDT powder to kill the body lice that spread a serious illness. At the same time, DDT was sprayed in houses to control house flies and the mosquitoes that carry malaria. In both cases the strongest insects survived the pesticides. The treatments were repeated. Each time, fewer and fewer insects died.

In 1948 a new chemical, **chlordane** (KLOR-dayn) was added to the DDT spray. The new spray worked well for two years. However, by 1950, **resistant** (ree-ZIS-tant) flies appeared: They were not killed by chlordane. By the end of that year, all the house flies and mosquitoes seemed to be resistant. By the end of 1951, six new chemicals had been tried. None worked for very long.

At the end of the treatment period the fly population was just as large as before. But the flies were now resistant to six chemicals. The environment was also loaded with the dangerous hard pesticides.

"SUPER-PESTS" DESTROYING CROPS

Insects that prey on farm crops are also showing resistance to chemical treatment. In 1960, 65 kinds of insects were known to be resistant. Now there are over 200. Among these are the codling moth, whose larvae makes apples "wormy," and the Colorado potato beetle. There are now six kinds of resistant insects that feed on cotton around the world.

FIGURE 22.19
The body louse is an insect that spreads the fatal disease typhus.

FIGURE 22.20
A Colorado potato beetle eats the leaves of potato plants.

The facts are clear. Chemical treatment of crops to prevent plant disease and crop destruction is sometimes effective. But resistant strains of insects often appear. More poisonous chemicals are then needed. The cost of controlling insects by pesticides then goes up. The new, expensive programs often work for a while. Then resistant insects appear again. In addition, a lot of poisonous chemicals have been added to the environment. So spraying programs often spread poison on the land, but don't do the job that they are supposed to.

Can you think of other ways to control insects that live on people or in their houses? Do you think that living with most insects is a better alternative than poisoning the environment?

Report 4. outwitting pests

Some scientists are trying to control insect pests without harming other living things. They have used two methods. They either try to trick the pests, or use organisms that prey on the pests.

IMITATING NATURAL PERFUMES

In some kinds of insects, the male finds the female during the mating season by following the smell of a chemical she releases. You may remember from Perceiving the World that many moths do this. Some flies do too. Scientists, knowing this fact, have made chemicals that work in the same way. Man-made perfumes have been produced that will attract male fruit flies and melon flies.

One way to use the perfume is to combine it with a poison. Then the male flies can be killed off and the population wiped out. Furthermore, when the poison-perfume combination is put on pieces of fiberboard, it attracts only the target insects. Other animals are less likely to eat it. That way the poison would not get spread over the land. This technique was tried out in 1960 on islands near Japan. One year later, over 99 per cent of the fruit fly population had been destroyed.

FIGURE 22.21

A Texas rancher is opening a box of sterilized screwworm flies. Boxes like this are dropped from planes. The boxes open when they hit the ground and release the sterile male flies.

INTRODUCING STERILE MALES

In this method huge numbers of a pest are raised in a laboratory. There they are given heavy doses of radiation. The radiation makes the males **sterile** (STER-uhl) or unable to produce offspring. Swarms of these sterile males are then taken to where the wild pest is doing damage.

The treated males from the laboratory mate with the wild females. Because the males are sterile, the eggs do not

346 What can we do about pests?

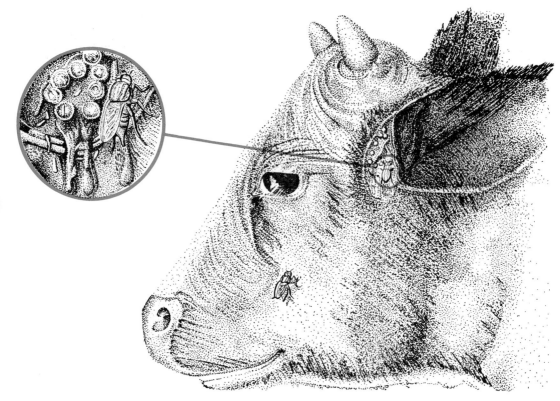

FIGURE 22.22
The female screwworm lays her eggs in cuts and scratches on livestock. The eggs hatch into maggots and may eat the animal alive in 10 days.

develop. This method eliminated the screwworm, a serious cattle pest, in Florida. It is also being tried in California to control a kind of bollworm that feeds on cotton plants.

PESTICIDES FROM BACTERIA

Some bacteria produce poisonous substances. One kind produces a poison that kills the caterpillars of the flour moth, cabbage butterfly, the root borer of bananas, and other insects. This poison is spread over plants. When the caterpillars eat the treated plants, they die. The difference between this poison and man-made pesticides is that the bacterial poison kills only certain insects. It is not harmful to any other organisms. Several companies are now using bacteria to produce this poison. Their scientists are testing ways to use it effectively.

SPREADING VIRUS DISEASES

In California, fields of **alfalfa** (al-FAL-fa), a hay crop, are now being sprayed with a virus. The virus gives alfalfa caterpillars a fatal disease. This saves the alfalfa, and it does not harm the livestock that are fed alfalfa.

In Canadian forests, scientists are attempting to control the pine sawfly, a pest of lumber trees, in a similar way. They are spraying the forests with another virus. This virus kills sawflies. The virus will become a permanent resident of the forests. Therefore, it should control the sawfly population over a long period of time.

No evidence has shown that these viruses can cause a disease in another animal.

INTRODUCING OTHER NATURAL ENEMIES

The use of other kinds of natural enemies to control harmful insects also shows great promise. In 1880 a scale insect threatened to wipe out the citrus trees in California. A small beetle that ate nothing but scale insects was imported. It has kept the scale insects under control. Recently, however, chemical sprays used to control other insects have killed

FIGURE 22.23
Cottony cushion scale of citrus trees is produced by tiny insects that feed on sap.

off this desirable beetle. The scale insects are less affected by the sprays. They are again threatening the citrus groves.

Canada is trying another natural method to exterminate sawflies. The shrew, a small animal like a mouse, was recently introduced to Newfoundland, Canada. The sawfly lives part of its life in the soft soil beneath the trees. Here it is often eaten by shrews. One shrew can eat as many as 800 sawflies each day. It is not yet known whether the shrews will help control the sawfly, but early results seem promising.

Why do you think that these methods seem to work? Can you think of any ways to get rid of pests in your home and yard without using poisonous chemicals?

Gathering data

You and your classmates have read and discussed some problems that result when man makes serious changes in his environment. Do these kinds of problems exist in your community or in your state? You could call your local conservation officer, Audubon Society, or newspaper to find out. Or, you might start looking for information in your school library.

Presenting the data

If you find information you believe everyone should know, how can you tell them about it? Here are some suggestions.

a. Make a set of color slides showing your data and prepare a talk to go with them. You can present this to other classes or to groups of interested adults.
b. Make large charts, graphs, or posters to display around the school or in prominent places in the community.
c. Prepare and distribute pamphlets which give the results of your study.
d. Put on a skit or play for other students in your school.

Whatever you choose, look for well-informed people to help you. That is, talk to the art teacher about posters or signs, to the school photography advisor or local photographer about pictures, or to the drama or English teacher about a play. Look for the person who has the special knowledge you need to make the best presentation.

FIGURE 22.24
These common insects help man by eating pests; a *lady bug beetle,* b *praying mantis,* c *dragonfly.*

a

b

c

Mastery Item 22-1

A pesticide poll

What is your opinion about the following statements? Choose the box that is closest to the way you feel. List the letters *a-i* on a piece of paper. Next to each letter, write down the box you chose. Give your reasons for choosing that box.

a. Man should try to get rid of all the pests that threaten or bother him.

strongly disagree	disagree	neutral	agree	strongly agree

b. Keeping down weeds and insect pests in your own yard is more important than the little bit of poison you add to the environment.

strongly disagree	disagree	neutral	agree	strongly agree

c. The effects of pesticides on man must be well understood because the United States Government has exact limits on how much pesticides are allowed in our food!

strongly disagree	disagree	neutral	agree	strongly agree

d. Farmers really shouldn't pay too much attention to all the effects of pesticides, because if the pesticides work, that is all that counts.

strongly disagree	disagree	neutral	agree	strongly agree

e. I wouldn't mind buying food that was slightly damaged by insects, if this is the result of growing crops free of pesticides.

strongly disagree	disagree	neutral	agree	strongly agree

f. Recently a manufacturer has introduced a new shelf paper for kitchen cupboards. It has a pesticide on it that keeps insects out of the cupboards. If I had ants in my cupboard, I would put this paper on my shelves.

| strongly disagree | disagree | neutral | agree | strongly agree |

g. Recently a city in Ohio was invaded by huge numbers of mosquitoes. Even children waiting for school buses were bitten. This community decided that the mosquitoes were coming from a nearby wildlife refuge which was quite swampy. To protect their children, the city decided to spray the swamp. The people who managed the refuge said that spraying could do more harm than good. How would you feel about this proposal?

| strongly disagree | disagree | neutral | agree | strongly agree |

h. If pesticides were really dangerous, the government would not allow them to be sold in supermarkets.

| strongly disagree | disagree | neutral | agree | strongly agree |

i. Passing laws to eliminate DDT will solve our pesticide problem. Other pesticides are not dangerous.

| strongly disagree | disagree | neutral | agree | strongly agree |

Key

Discuss your answers with your classmates and teacher.

FIGURE 23.1
A London street can barely be seen during a recent smog. The glow in the upper right corner is the sun!

Investigation

23   Why is clean air important?

On December 4, 1952, a thick fog rolled in over London, England. Hardly anyone paid any attention at first, because the city is famous for its "pea soup" fogs. However, this fog did not burn off or blow away. It combined with smoke to form smog. As moisture, soot, and smoke collected, they turned the smog from white, to brown, to black. Buses traveled at two miles an hour. Someone had to walk ahead to direct the drivers. Firemen had to walk in front of their engines. A performance of the opera was halted in Act I because the smog seeping into the theatre was so thick. The singers could not see the conductor. The audience could not see the stage. After three days of smog, the visibility was reduced to inches.

People having trouble breathing packed the hospitals. For five days the smog smothered London. The British Committee on Air Pollution estimated that there were 4,000 more deaths than usual. During the next two months, there were 8,000 more deaths than usual. Scientists suspected these, too, were caused by the killer smog.

Donora, Pennsylvania, was hit by a smog on October 26, 1948. This town is located near the bottom of a steep valley. This time, the fumes from a steel factory, a sulfuric acid plant, and a chemical plant settled down on the town. The atmosphere became horribly polluted with industrial wastes. Street lights had to be turned on during the daytime. Some 5,000 people became ill, and 20 died from the effects of the smog.

FIGURE 23.2
A student's view of the future.

The Los Angeles smog, as serious as it is, has even become famous as a source of jokes. One T.V. comedian said that people in Los Angeles wake up in the morning to the sound of birds' coughing!

Just what makes air polluted? How can you measure the amount of pollution in the air in your community? How much pollution do you breathe every day? How does air pollution harm living things? You will investigate these questions in this Investigation.

Problem 23-1

How much pollution falls on your house?

Two basic things make up air pollution. They are *particles* (often called particulates) and *gases*. Smoke contains both. The particles are soot and dust. They tend to settle out of the air. However, some very small particles float in the air

FIGURE 23.3
Next they'll put people in bags and leave the pollution outside.

until they are washed out by rain and snow. Or they may combine with other substances. Scientists also know that particles in the air can collect moisture. Weather records show that more rain and snow fall in smoky areas.

Where there are particles, there are usually polluting gases. For example, industries and cars give off both. The gases are invisible, and they are often present in very small quantities. Complicated equipment is needed to detect gases and measure them. Therefore, you will measure particles of air pollution in this Problem.

Your class is going to collect information about how polluted the air is in your community. Each person will be responsible for gathering data in this survey.

Materials
map of the community
gallon jar
distilled water
saucepan
hot plate or burner
balance

Gathering data

The class should look at the map and decide where to sample pollution. Try to cover as much of the community as possible. You and your group will collect samples. You will measure the total mass of pollution that settles down on a particular area. It could be the area you live in.

To find the mass of the particles that fall on your community will take about a month. The easiest way is to collect the particles in jars of water. Then you can evaporate the water and weigh the particles that are left. These directions may help you gather reliable data.

a. Pour about one quart of water into each sampling jar. Mark the water level with a waterproof marker so it will not wash off in the rain. Why do you think the experiment requires water in the jar? Why do you think it requires distilled water?
b. Label the jar with your name, date, and location. You might like to identify your jar as part of a pollution experiment from your school.
c. If possible, set each jar about the same distance from the ground. Five feet is ideal. Let the jar sit undisturbed for 30 days. Go and look at it every three or four days. Add more distilled water to keep it from drying out.
d. Bring the jar to school. Cover it so you don't spill any water on the way.
e. Accurately weigh the pan you will use to evaporate the water. Pour the water into the pan. Remove large objects such as leaves or insects. They aren't air pollution. Rinse the jar with distilled water and pour the rinse water into the pan, too.
f. Evaporate the water slowly. Use a system the teacher has approved. Slow evaporation is safer and will give more accurate results.
g. When the water has evaporated, let the pan cool off. Then weigh it again. Is the mass of the pan more the second time? Any difference in mass should come from particles that fell into the jar.

Recording data

Record the location of your jar on the community map. Also record any differences in mass. Did you make any other observations?

Why is clean air important?   357

FIGURE 23.4
How many sources of air pollution like this one can you find in your community?

Analyzing data

Now you know the mass of the particles that fell in your jar. How much fell in the neighborhood of the jar? To find out you have to do some arithmetic.

a. Measure the distance across the mouth of the jar, the diameter. Suppose, for example, that this distance is 20 centimeters.

b. Take one-half this number (10) to find the radius of the jar's mouth. Multiply the radius by itself (10 × 10), then by 3.14. This gives you the area of the jar's mouth in square centimeters (314 cm²). The formula is:

$$\frac{\text{diameter (cm)}}{2} \times \frac{\text{diameter (cm)}}{2} \times 3.14 = \text{area (cm}^2\text{)}$$

c. Divide the mass of the particles by the area of the jar's mouth:

$$\frac{\text{mass of particles (mg)}}{\text{area of jar's mouth (cm}^2\text{)}}$$

d. Multiply this number by 28.6. Now you know the number of tons of particles that fall on each square mile of your community in 30 days.

Did the same number of tons of pollution fall on each area in your community? How does the data from Problem 23-2 compare with this data? Can you make a hypothesis to explain the data? How could you test your hypothesis?

Problem 23-2

What do airborne particles look like?

As you set up Problem 23-1, you may have thought of some new questions. Here are some questions to answer in this Problem. How big are pollution particles? Are there different kinds? Can you find out where they come from? Is the number of particles collected the same each day of the week? Collecting data to answer these questions is easy.

Why is clean air important?

Materials

map of the community   microscope
glass slides   clothespins
clear plastic millimeter ruler

Gathering data

Particles can be collected on a glass slide. Coat most of one side with a very thin layer of vasoline. Leave a place to pick it up. When this slide is exposed to air, particles will stick to it. Later you can observe these particles under a microscope.

You are responsible for collecting in one area. The class should agree on what areas to sample. Number the sampling areas on the community map. Take your coated slide to your sampling area.

Hang it on a string or clothesline with a spring-type clothespin. (See Figure 23.5.) Or lay it on a flat surface. Collect the slide after 24 hours. Mark down on it the date and the location.

FIGURE 23.5

Here is one way to sample the particles of air pollution in your community. What other ways could you use?

Observe your slide through a microscope. Are the particles different sizes and shapes? Are they all the same color?

You should compare your sample of air pollution with others. To do this you need to know the average number of particles in an area of your slide. Here is how to find the average number: Without looking through the eyepiece, center the slide on the stage. Focus the microscope and count the number of particles you can see. The area of the slide that you can see at one time is called the *field*. Record the number of particles in the field.

Now move the slide a short distance. Count and record the number of particles in this new field. Repeat this procedure several times. Make sure that you are looking at a different part of the slide each time. Now find the average number of particles in a field. Find the average by dividing the total number of particles by the number of fields you observed. Record this number.

You could compare your number with a classmate's. But the field of his microscope may not be the same size as yours. Therefore, you need to know the number of particles on a square millimeter of your slide.

a. First, measure the width of the low-power field of your microscope. Lay a clear plastic, millimeter ruler across the microscope stage as shown in Figure 23.6 Place it so that one of the millimeter lines can just be seen at the left edge of the field. Make sure that the ruler crosses the center of the field. You will be able to see a second line in the field.
b. The distance between the markings is one millimeter. How many millimeters wide is your field? Estimate it. This total distance is the diameter.
c. Substitute this number in the formula below to find the area of your field.

$$\frac{\text{diameter (mm)}}{2} \times \frac{\text{diameter (mm)}}{2} \times 3.14 = \text{area (mm}^2\text{)}$$

d. Now, divide the average number of particles per field by the area of the field. This will give you the average number of particles per square millimeter (particles/mm^2)! Now you can better compare your slide with your classmates' slides.

FIGURE 23.6

Use a clear plastic ruler to measure the diameter of your microscope's field. This will help you compare your data with your classmates'.

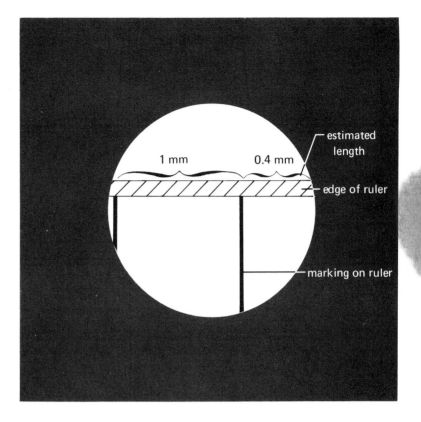

Recording data

You should have already described the particles you saw and recorded your particle counts.

Now you can record your observations on the map of the community. Put down the particles/mm^2 and the date you exposed the slide.

Analyzing data

Where did you find the most particles? Did you find more on weekdays than weekends? Can you explain the differences? Do the residential areas, highways and railroad tracks, industries, and dumps make different amounts of air pollution?

From the size, color, and shape of the particles, can you infer how many different places they came from?

Present a report of your findings to the rest of the school and to the P.T.A. Perhaps the art teacher can suggest how to illustrate your report. Are you bothered by your data?

362   *Why is clean air important?*

Does it raise questions you would like to ask the people that govern your community? One way to get answers is to address these questions to the local newspapers. You can write letters to the editors and ask for answers to your questions.

Problem 23-3

How much dirt do you inhale?

Clean air is a mixture of invisible, odorless, and tasteless gases. You know that the air also contains man-made gases and particles. Few of us would drink a glass of milk or water

FIGURE 23.7
This photograph shows some sandstone pillars in the middle of a city. Acids in the air have eaten them away. One has been replaced with a cement pillar.

which we knew contained these substances. Does it bother you to know that you inhale them?

Adults need about four pounds of water a day to stay alive. They need about two and three-quarters pounds of food. How much air do you need each day? How much soot and polluting gases are in that air?

You can design simple ways to measure the amount of air you breathe in a day. Then you can calculate how much of that air might be pollution.

Does everyone in the class inhale the same amount? Do the girls inhale more pollution than boys? Does exercise affect the amount of dirt you inhale? Do heavy people breathe in more air than thin people? After finding out how much air everyone breathes, you can think of ways to answer these questions.

Materials

The materials in Figure 23.8 can be used to measure the volume of a breath of air. Perhaps you can think of other materials.

FIGURE 23.8
You can use some or all of these materials to measure the volume of a breath of air.

Gathering data

Design a quick and simple way to measure the approximate volume of air in a normal breath. Use some (or all) of the materials in Figure 23.8. Decide on the method your group will use. Then ask your teacher for the materials you need. Will you need more than one sample?

Now estimate how many times you breathe in an hour, then how many times in a 24-hour day. Calculate the volume of all the air you would breathe in a day. Then use Figure 23.9 to find out how much pollution is in that amount of air. Use the data from the city nearest yours.

FIGURE 23.9
Average air pollution levels in American cities

City	Micrograms* of particles/m^3 of air	Micrograms of sulfur dioxide/m^3 of air
Los Angeles, California	119	35**
San Francisco, California	68	15
Denver, Colorado	126	18
Hartford, Connecticut	90	62
New Haven, Connecticut	110	101
Wilmington, Delaware	124	111
Washington, D.C.	77	90
Chicago, Illinois	124	221
East Chicago, Indiana	186	107
Indianapolis, Indiana	154	54
New Albany, Indiana	129	38
Des Moines, Iowa	123	13
Dubuque, Iowa	132	16
Covington, Kentucky	112	35

*Microgram is one-millionth of a gram.
**Data for Los Angeles was not available. This figure is for Long Beach, California.

City	Micrograms* of particles/m³ of air	Micrograms of sulfur dioxide/m³ of air
Baltimore, Maryland	146	107
Boston, Massachusetts	149	30
Detroit, Michigan	161	16
Minneapolis, Minnesota	87	44
Kansas City, Missouri	116	12
St. Louis, Missouri	143	132
Glassboro, New Jersey	65	43
Newark, New Jersey	96	174
Buffalo, New York	139	25
New York City, New York	134	346
Cincinnati, Ohio	154	44
Cleveland, Ohio	116	78
Oklahoma City, Oklahoma	107	10
Portland, Oregon	75	22
Pittsburgh, Pennsylvania	151	93
Bayamon, Puerto Rico	128	8
Guayanilla, Puerto Rico	36	4
Providence, Rhode Island	121	125
Chattanooga, Tennessee	140	34
Nashville, Tennessee	115	29
El Paso, Texas	193	63
Salt Lake City, Utah	93	20
Seattle, Washington	76	35
Charleston, West Virginia	226	29
Milwaukee, Wisconsin	150	28

Source: U. S. Dept. of Health, Education, and Welfare. *Air Quality Data.* Durham, North Carolina, 1966.

What factors affect how much air you inhale? Make a hypothesis. Does height? Does exercise? Do boys inhale more than girls? What other factors might be important? Test your hypothesis. If necessary, design an experiment to get any extra data you need.

Recording data

Keep a complete record of all the procedures you use for this Problem. For example, record your method for finding how much air you breathe in a day. Write down the hypothesis you are testing and the data you collect. Describe any experiments you performed.

You can tell that you have a complete record when a classmate can use your descriptions and perform the investigation exactly as you did.

Analyzing data

Does your data support your hypothesis? Make a short report to the class.

Discuss your results with your classmates. What people inhale the most air? What factors seem to cause people to breathe in more air? Why? If you moved to Seattle would you breathe in less pollution? If you moved to New York would you breathe in more?

Problem 23-4

Can air pollution harm people?

Figure 23.10 is a diagram of the human respiratory system. If you know how it works, then you can understand how pollution might affect respiration.

The larger air passages are lined with cells which produce a sticky substance called **mucus** (MYOO-cuss). Other cells that line these passages have hairlike parts called **cilia** (SILL-ee-uh). These cilia wave back and forth. They paddle the mucus up through the air passages to the throat, where it is swallowed. The stream of mucus is like a conveyor belt. It carries particles out of the respiratory system. Almost 90 per cent of the particles inhaled are caught in the mucus.

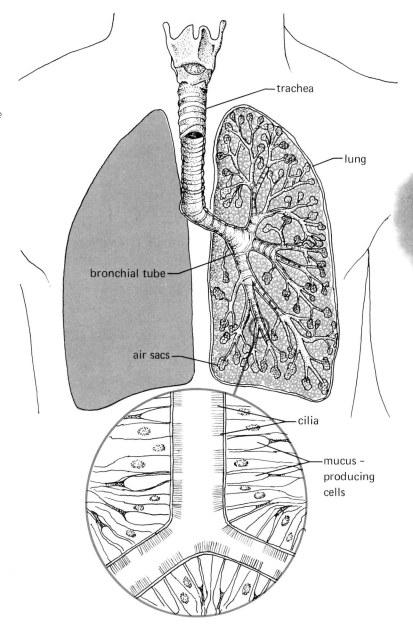

FIGURE 23.10
In the human respiratory system, the air passages branch many times, and finally end in clusters of tiny air sacs which make up the lungs. The inset shows that the air passages are lined with a cleaning system.

This cleaning system can be damaged. For example, sulfur dioxide, smoke, and certain types of smog can damage the cells lining the air passages. Then so much mucus is produced that it clogs up the lungs. Pollution builds up in the larger passages: One coughs, wheezes, and chokes.

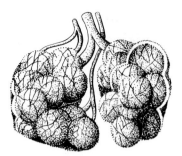
normal air sacs

air sacs affected by emphysema

FIGURE 23.11
How does emphysema affect the lungs? What differences can you find between the two sets of air sacs above?

The smaller air passages can also be damaged. The large air passages into the lungs branch and divide. Eventually they end in millions of air sacs. Here gas exchange takes place. Oxygen goes into the blood, and carbon dioxide comes out. These tiny sacs are smaller than the head of a pin. The delicate walls of the sacs don't have any protective mucus and cilia. Therefore, they are easily damaged by polluted air.

Life insurance companies say that more and more Americans suffer from severe respiratory diseases each year. **Emphysema** (em-fuh-SEE-muh), a crippling disease, is becoming common.

When someone is sick with emphysema, his air sacs become stretched. See Figure 23.11. The stretched sacs exchange less carbon dioxide and oxygen than normal air sacs. As the disease gets worse, breathing becomes difficult. The tiny blood vessels in the air sacs shrink or die. The lungs also become more easily infected. The exact causes of emphysema are unknown. Damage from sulfur dioxide may be one cause. Almost certainly, cigarette smoking is another. Some patients suffering from this disease show improvement when they are given unpolluted air to breathe. Can you see why this might be so?

Scientists know polluted air can harm some animals. But there isn't as much data on how polluted air might affect

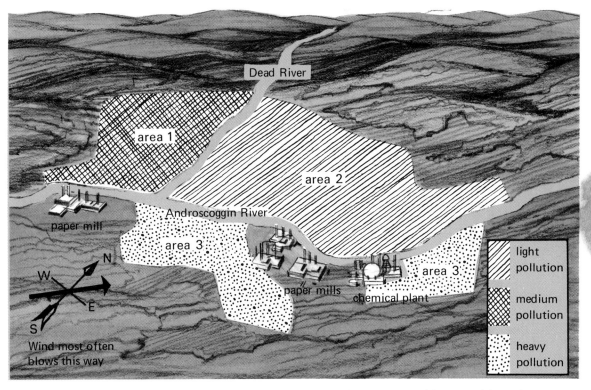

FIGURE 23.12
This is a pollution map of Berlin, N. H., made in 1961. It shows which way the wind usually blows and where the big industries are located. Knowing these two things, could you have predicted which area would have the worst air pollution?

people. Short exposure may cause only watery eyes and mild coughing. It is also possible that air pollution might cause diseases. Or it might cause mild diseases to become more serious. You be the judge. You can analyze some data collected by doctors* in the New Hampshire community of Berlin in 1961.

Analyzing data

Doctors wanted to know if the air pollution from factories caused illness. They tried to find out if more people had lung diseases in areas with heavy pollution. The air in the town was surveyed. Figure 23.12 shows there were areas of light, medium, and heavy pollution. The doctors then investigated how many people had breathing problems in each

*Ferris, Benjamin G., Jr. and Anderson, Donald O. 1962. The Prevalence of Chronic Respiratory Disease in a New Hampshire Town. **Am. Rev. of Respiratory Diseases.** 86:165–177.

Why is clean air important?

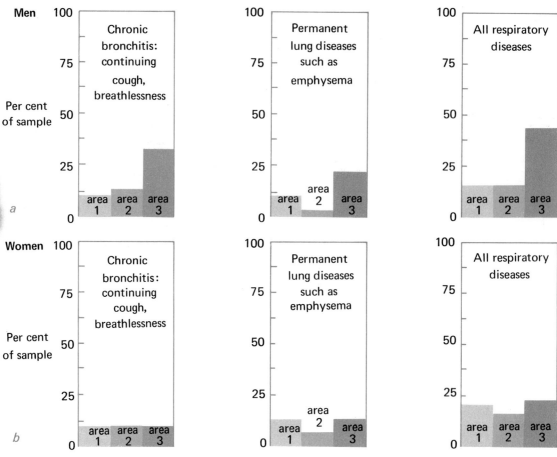

FIGURE 23.13
a. The bar graphs show what per cent of men in Berlin, N. H. had a respiratory disease. They are classified according to where they live. None of the men were smokers.
b. The graphes show the amount of respiratory disease in women in Berlin, according to which part of town they live in. Where would you want to live if you were moving to Berlin?

area. Where would you predict breathing problems were worst? Figure 23.13 gives some of the doctors' data. Suppose you and your family were moving to Berlin, New Hampshire. What part of town would you want to live in? Discuss the data and your conclusions with your classmates. Are men or women more likely to get a respiratory disease? What else would you like to know about the conditions in Berlin, New Hampshire, before moving there?

Problem 23-5

How does sulfur dioxide affect plants?

Air pollution makes it difficult to grow plants in or near a city. Only a few kinds of trees can survive in city air. Even in areas far from cities, smoke and smog are damaging trees. The smog from Los Angeles is killing 161,000 acres of pine trees in a national forest. Some of these trees are 115 miles away from Los Angeles! Christmas tree growers in Maryland may lose as many as 300,000 trees each year because of the smoke from power plants. Farmers in many parts of the United States are having trouble producing crops. For example, smog harms lettuce, beets, spinach, and orange trees.

What happens to plants? Which parts suffer first? How long does it take to do damage? You can do a simple experiment to answer these questions. You will use a common source of air pollution, sulfur dioxide. It comes from burning coal and oil. Power plants and heating systems of homes and apartment buildings put a lot of it into the air.

Materials

at least 2 plants   a small container   tape
large plastic bag for each plant. If the bags have holes, tape them shut.
2 g sodium sulfite (a chemical made of sodium, sulfur, and oxygen)
2 ml of 5 per cent sulfuric acid. *The teacher will bring the acid when you are ready.*

Gathering data

A simple way to test the effect of air pollutants on plants is shown in Figure 23.14, on the next page.

You can make sulfur dioxide by adding sulfuric acid to sodium sulfite. CAUTION—THESE TWO CHEMICALS SHOULD BE MIXED *ONLY* IN A CLOSED CONTAINER. DO NOT BREATHE THE SULFUR DIOXIDE FUMES.

Put a plant in the bag. Set it down on the desk. Then put in the container of sodium sulfite. Ask your teacher to add the acid to the sodium sulfite. After the acid is added, quickly seal the bag with tape. What is your control?

FIGURE 23.14
Your teacher will help you add acid to sodium sulfite crystals. This is a safe way to find out how sulfur dioxide affects plants.

Recording data
Record the appearance of each of your plants before the experiment. You could draw or photograph them. Otherwise, describe such things as the color of the leaves and any wilting you see. Observe and record the appearance of each plant for two or three days during the experiment.

Analyzing data
Combine your data with data of classmates who used the same kind of plant. What conclusions can you draw from everyone's data?

Does the kind of plant you used make any difference? Are plants with thick, waxy leaves affected the same way as plants with long, thin leaves? Are young plants more easily damaged than older plants? Do you think smog would affect house plants?

Why is clean air important?

Mastery Item 23-1

Planning a local pollution study

Suppose you were moving to Berlin, New Hampshire, as it suggests in Problem 23-2. How could you bring the map in Figure 23.12 up to date? What other data would you collect to describe the air pollution problem?

Key

You could set up a sampling program that would do most of these things:

a. Cover the entire community.
b. Map the distribution of pollution.
c. Gather data on amounts of pollution.
d. Identify sources of pollution.

You may have had other ideas. Discuss your answer with your teacher.

Mastery Item 23-2

Inhaling pollution

Look at Figure 23.15 on this page and the next two. Who do you think is breathing in the most air pollution? List the letters of the pictures, in order, from most to least.

FIGURE 23.15

Which of the people shown on this page and the next two is breathing in the most air pollution? How much do you think you are breathing in right now compared to these people?

a

Why is clean air important?

b c d

FIGURE 23.15 cont.

Why is clean air important?

e

f g

Key

C and f are probably breathing in the most. Then comes a, followed by g. B and e are probably breathing in little air pollution. D would be the least. All these answers depend on inferences about the quality of air in each scene. If your list is different, discuss your reasons with your teacher.

Mastery Item 23-3

Do automobile exhausts harm corn?

Some farmers along large highways are worried about the health of their corn crops. They have read that smog kills trees and crops around Los Angeles. Some of them think that exhaust fumes from cars and trucks are damaging the corn.

How could you test their hypothesis? Design an experiment and write down the procedures you would follow. What evidence of plant damage would you especially look for?

Key

If you tested the hypothesis in a laboratory, you would need:

a. a method for exposing corn plants to about the same amount of gases that plants in the field get;
b. control plants growing in clean air;
c. a way to measure results, such as observations of height, rate of growth, leaf color, number of ears of corn produced, size of the ears, or the quality and size of the corn kernels.

You might have thought of another way, for example, observing corn plants in the field. Your design should include b and c above. Discuss your idea with your teacher.

Mastery Item 23-4

Does smoking cause lung cancer?

Not all air pollution comes out of exhaust pipes or brick chimneys. Cigarette smoke is a common kind of air pollution that requires only a human chimney.

American Indians taught the English how to smoke in the seventeenth century. The tobacco plant did not grow in England or Europe. At that time Englishmen both praised and blamed the new import.

There was no scientific evidence to support either side. Then in 1939, cancer was first linked to tobacco. Tars from

FIGURE 23.16
Sir Walter Raleigh was one of the first Englishmen to smoke. Which of the quotations would you agree with? Do you think that the argument about the value of tobacco has finally come to an end?

tobacco were painted on the backs of rabbits. They caused cancer of the skin. But this evidence did not say that people get skin cancer from tars. Nor did it say that tars cause lung cancer.

However, there is data from public health records about smoking and lung cancer in humans. Many doctors, including the Surgeon General of the United States, are convinced that this data indicates that smoking cigarettes can cause lung cancer. Some of this data on smoking is presented in Figures 23.17 and 23.18 on the next page. Your task is to analyze the data. Decide whether it supports the hypothesis that lung cancer is associated with smoking.

FIGURE 23.17
This data relates the amount of smoking to death from lung cancer. Can you find a pattern?

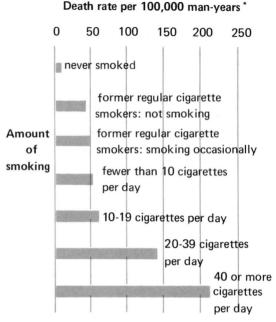

*A man-year is data on one man for one year.

FIGURE 23.18
This data shows death rates from lung cancer. The data includes smokers and nonsmokers in different environments. Who is most likely to die of lung cancer? Who is least likely?

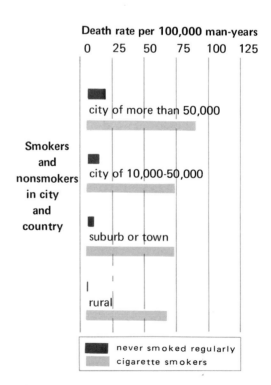

Why is clean air important?

Figure 23.17 shows that more smokers die from lung cancer than nonsmokers. Also, as smoking becomes heavier, the death rate increases.

Figure 23.18 shows that in cities of any size more smokers than nonsmokers die of lung cancer. The figure also shows that people living in clean air who do not smoke are least likely to get lung cancer.

Key

Mastery Item 23-5

Stop smoking!

Consider what you now know about the effects of dirty air. Write a letter to someone you know who smokes and, using the data you have, try to convince him to stop.

The letter might include information on:

a. How pollution affects the linings of the respiratory system and the air sacs in the lungs.
b. The special danger of smoking right after exercising: so much more pollution is inhaled.
c. The relationship between lung cancer and smoking.

Key

FIGURE 24.1
Do you think the students in this anti-pollution demonstration are exaggerating about the seriousness of the situation?

Investigation

24 The pollution game

If our country is in a mess from many kinds of pollution, why doesn't someone do something about it? If a person wants to try to make his environment a better place to live, where does he start?

What kinds of problems or situations make it difficult to clean up pollution? When you play "The Pollution Game," you will deal with some of these problems. You will see how the actions of individuals and organizations affect others. You will also see some ways to control pollution.

Your teacher will show you the different parts of the game and then go over the rules with you.

How to play

1. Each team gets one set of "The Pollution Game." A team consists of four players and a banker. The game can be played by four players, if one player also acts as banker. The banker is a cashier. He does not vote or own property or a car.
2. Team players roll the dice to determine the ownership of property and the order of play. The highest number picks a colored set of properties and makes the first move. (All the properties in a set are the same color.) The next highest number picks another set of properties, and so on. The lowest number plays the banker.
3. When the game begins, each player gets $3,500 and two **Election Cards** from the banker. Each player receives $500 each time he passes **Start**. To advance around the board, a player throws the dice and moves clockwise the number of spaces shown on the dice.

FIGURE 24.2
Cars cause 60 to 70 per cent of all air pollution in cities. How could you solve this tremendous problem?

4. Each player collects rent or fees from the other players that land on his property. The rates are given on the board. Some fees change as the **Air** and **Water Pollution Indexes** change.
5. To begin, set the **Air Pollution Index** and the **Water Pollution Index** at 10. As certain businesses make money during the game, they pollute the air, or water, or both. Players' cars also cause air pollution. The changes in pollution levels are kept on the Pollution Indexes by the players. If either Index reaches the **Lethal limit,** all members of the team lose. The environment won't support life. An Index does not change if a player lands on his own property.
6. When a player passes Speedy Auto Company, he must get his car tuned-up and his pollution control device serviced. You can choose from three kinds of tune-ups. The best tune-up costs the most, but prevents the most pollution.
7. To get money to stay in the game, a player may sell some of his properties. The buyer and seller have to agree on a price. Any player who loses all of his properties is out of the game. If a property is unowned, no rents or fees are collected on it. Players can keep a written record of any sales of property.
8. When a player lands on **Take a Chance,** he must take the top card and read the instructions out loud. All players affected must follow the instructions before the next player takes his turn. If all of the cards have been used once, shuffle them up and begin again with the top card. Unless the card says otherwise, Take a Chance fees go to the bank.
9. If a player wants to reduce the pollution, he can call for an election. Any player can call for an election when he passes start. At the same time he must give one of his election cards to the banker. Since each player has two election cards, the team has only eight chances to change the rules.

   During an election the players select the proposal they would like to adopt. A majority vote is necessary to pass a proposal. Individual players are free to vote as they see fit.

   The first time a team plays the game, all election proposals must be chosen from the Ballot. The second time the team can make up its own ballot.

10. The game lasts either one or two class periods. This should be agreed on before play begins. At the end of that time, the team that has the lowest air and water pollution indexes wins. The individual winner in the class is the player who has the most money. All players on a team lose if they pollute to the lethal limit.

Election Ballot

Proposal 1

Charge all properties shown on the Mill City map for city pollution control. Players must add up the current rents and fees for these properties and pay half the amount to the bank. This will lower the Air and Water Pollution Indexes by 15 points. (This Proposal can be voted more than once.)

Proposal 2

Each player invests $200 for research on ways to control air and water pollution. This investment in research may pay off in the future. Watch the Take a Chance cards. (This proposal can be voted more than once.)

Proposal 3

Pass a Mill City law against adding poisonous wastes to the Little Muddy River. Valley Steel and Gordon Paper Co. must each pay to the bank $800 to install new equipment. Each company will then pollute the water at the lower rate of 5 points (it was 10) for each $500 earned.

Proposal 4

Pass a law to make Atomic Electric Company cool the water it discharges into Clear Lake. It will cost Atomic Electric $1,200. The Company's pollution rate will then go down to 2 points (it was 4). But customers will have to pay $600 now, instead of $400.

Proposal 5

Build a new waste-disposal plant for Mill City. The new plant would dispose of all garbage and trash without producing a lot of smoke or gases to pollute the air. Each player must contribute $100. The Air Pollution Index will go down 15 points.

The pollution game

Proposal 6

Pass a Mill City law to limit the amount of smoke and gases coming from chimneys. Valley Steel, Gordon Paper, and Progressive Chemical Company will have to pay the bank $800 each to install air filters on their chimneys. Each company will then pollute the air at the lower rate of 5 points (it was 10) for each $500 earned.

Proposal 7

Pass a law to require Mill City Power Company to burn only a high-grade, smokeless coal. The air pollution rate will then go down to 2 points (it was 10). To pay for the more expensive coal, the company will now charge $600, instead of $400.

FIGURE 24.3

Trash and litter—especially materials that do not rot—add to the pollution problem.

Proposal 8

Pass a city ordinance requiring all players to get a "super tune-up" each turn around the board. This will keep air pollution from automobiles as low as possible.

Proposal 9

Build a mass transit system. This will reduce automobile traffic and therefore reduce air pollution. Construction will cost each player $600. Automobile tune-ups will not be required each round.

Proposal 10

Build an improved sewage treatment plant for Mill City. The new plant would remove solids, harmful bacteria, and most phosphates and nitrates from city sewage. Each player must contribute $150. The Water Pollution Index will go down 15 points.

Players should keep a written record of the proposals they adopt. That will remind players of changes in rents and pollution rates. Use a form like the one shown in Figure 24.4.

FIGURE 24.4
Here is a sample ballot record sheet. Use one like this to keep a record of the proposals your team has adopted.

Proposal adopted	Company affected	Old pollution rate	New pollution rate
#4	Atomic Elec.	4 pts. / $400	2 pts. / $600

FIGURE 24.5
Most of the environmental problems we have today were caused by man. Can man correct these problems, or will the problems get worse and worse?

Appendix

Part A

Observing with the microscope

This section will help you learn to observe with a microscope when you want to or need to. You will probably be able to learn by yourself. But your teacher will help you when you ask.

You and some other students will probably be using the same microscope. Taking care of it is important. A microscope in good shape is easy and pleasant to use.

Materials
microscope
several microscope slides and cover slips
lens paper   medicine dropper
small amount of water forceps

Making a wet mount
Find something you want to observe. It might be hair, aquarium water, a leaf, or just about anything. Select a tiny bit of the substance *thin enough for light to pass through*. Then make a slide using the procedures in Figures 1–4. This is called a wet mount because the substance is placed in a drop of water.

Observing with the microscope

FIGURE 1

Set a glass slide down on your desk or table. The object that you want to observe may be picked up and put on the slide with forceps.

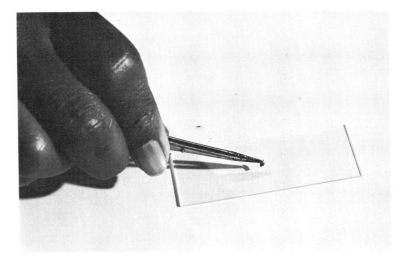

FIGURE 2

Put one or two drops of water onto the object, enough to cover it. If you are looking at pond water or another liquid, put one drop of it on the slide.

FIGURE 3

Carefully pick up a coverslip. Put one edge of it beside the drop of water on the slide. Slowly lower the coverslip. The water should flow under it. This method will help you avoid getting lots of bubbles under your coverslip. The more slowly you lower the coverslip, the more luck you should have.

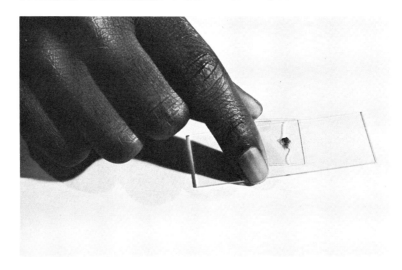

FIGURE 4
Now you have something ready to look at with a microscope.

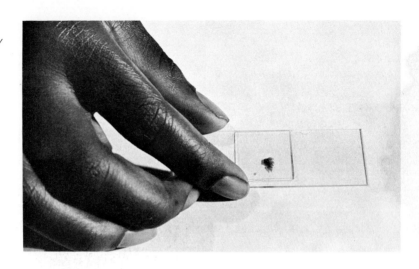

Using the microscope

Set aside your wet mount and get a microscope to use. In some ways your microscope may be different from the one in the pictures. But it will probably be similar enough for you to follow the procedures below.

FIGURE 5
Carry the microscope upright with both hands. Hold the microscope by its arm and base.

FIGURE 6
Always set the microscope away from the edge of the table and keep it there.

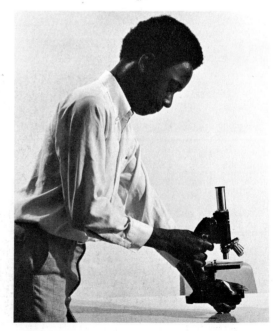

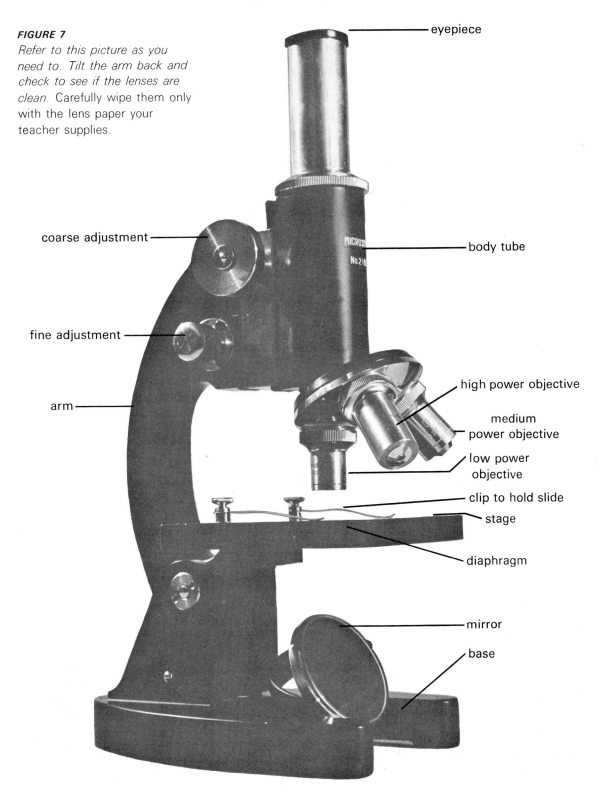

FIGURE 7
Refer to this picture as you need to. Tilt the arm back and check to see if the lenses are clean. Carefully wipe them only with the lens paper your teacher supplies.

FIGURE 8
Make sure the lowest power lens (usually 5x or 10x) is under the body tube. Turn on the light under the stage. If your microscope has a mirror instead, turn the curved side up. Keep the microscope out of direct sunlight. *Look through the eyepiece and adjust the mirror to reflect light up through the hole in the stage. You should see a bright circle of light. If you move the microscope you will need to readjust the mirror to get the best lighting.*

FIGURE 9
Raise the body tube with the coarse adjustment. The low power objective lens should be at least one inch above the stage. Set the slide so that the substance you will observe is above the center of the hole in the stage. Hold the slide in place with the clips.

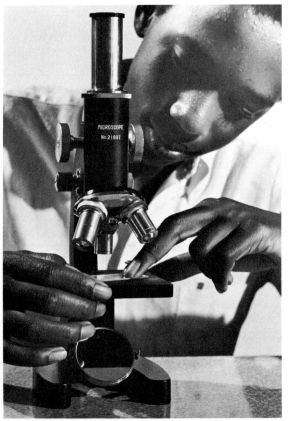

Observing with the microscope

FIGURE 10
Lower the body tube as you watch from the side. Stop with the lens about one-fourth inch above the slide.

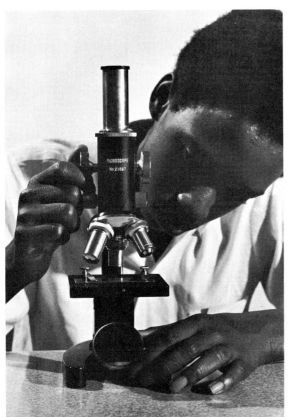

FIGURE 11
Look through the eyepiece. *Slowly* raise *the body tube with the coarse adjustment until the substance on the slide comes into view.* If you miss it, slowly repeat the procedures in Figure 10 and this Figure. Also, check to see if the substance is centered directly above the hole in the stage. Be sure to raise the tube to get the slide in focus. If you focus downward, you can break the slide and damage the microscope lens.

FIGURE 12
Use the fine adjustment to bring the substance into sharper focus. Move the slide around slowly to observe all of the substance. Turn the fine adjustment up and down just a little as you move the slide. Adjust the light with the **diaphragm** (DY-uh-fram) until you see the substance as clearly as possible.

FIGURE 13
To increase the magnification, rotate the next higher power lens under the body tube. You don't need to move the body tube. The substance should automatically be in focus.

Now adjust the focus with only the fine adjustment as you move the slide around. The higher power lenses don't let as much light through as the low power lens. Therefore, you may need to adjust the mirror to get more light or make another slide with less material.

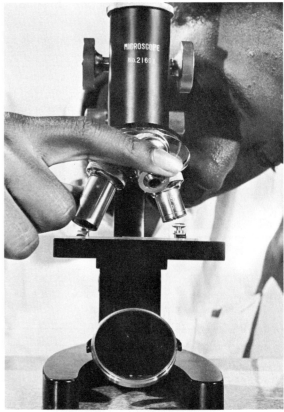

When you have finished observing, rotate the low power lens back under the body tube. If you want to look at another slide, repeat the procedures starting with Figure 9. If not, clean and dry the slide and cover slips.

Before putting the microscope away, always remember to check these points.

a. The low power lens is under the body tube.
b. The lens is lowered to about one-half inch from the stage.
c. The body tube is upright.
d. The diaphragm is wide open.
e. The lenses and stage are clean and dry.

Hold the microscope upright by its base and arm as you carry it.

Part B

How to graph data

This program of instruction will help you learn how to construct a graph. It consists of thirty numbered parts. Each part is called a "frame." Each frame gives you some information or a technique that will help you understand the contents of the following frames.

To begin, cover everything on the page below frame number one with a sheet of paper. Read the contents of the frame and answer the questions on a separate sheet of paper. Now uncover the printed answer to frame one. If your answer is correct, go on to frame two and so on. If your answer is not correct, reread frame one to discover why. Correct your answer, then go on. If you have any problems that you cannot solve, ask your teacher for help.

1.

The diagram on the right shows two lines that make the border of a graph. The border has two parts. They are the _____ and the _____.

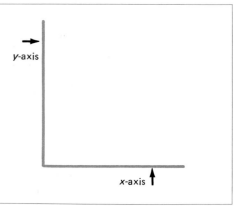

2.

| **Answer 1.** y-axis, x-axis |

The arrow in the diagram on the right points to the _____-axis.

3.

Answer 2. y

The arrow in the diagram on the right points to the _____-axis.

4.

Answer 3. x

Label each axis in the diagram on the right with its proper name. Write on your sheet of paper.

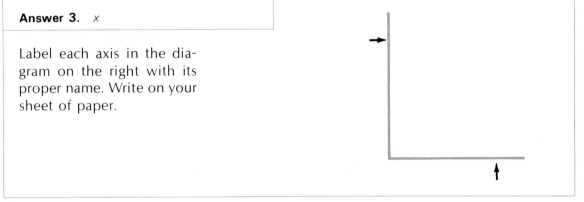

5.

Answer 4. The vertical line is the y-axis; the horizontal line is the x-axis.

Table 1 on the next page contains data collected in an experiment on goldfish. The x-axis in the diagram on the right is labeled with one of the variables, "Heart temperature, °F." One of the variables is used to label the _____-axis.

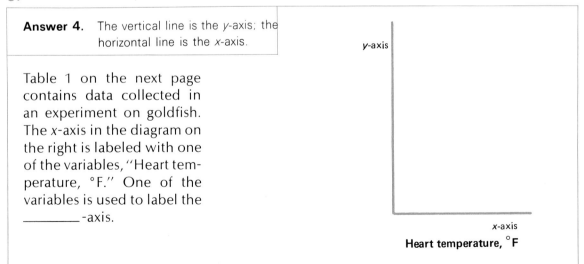

TABLE 1
The relationship of the number of heartbeats of a fish to heart temperature. This data was obtained by increasing the water temperature and by measuring the heart rate with an instrument. Since fish are cold-blooded, the temperature of their hearts is controlled mostly by the water temperature.

Heart temperature, °F	40	50	60	70	80
Heartbeats per minute	5	15	26	35	43

6.

> **Answer 5.** x
>
> The x-axis is commonly labeled with the variable that the experimenter controls directly. The experimenter directly controlled which variable in Table 1?

7.

> **Answer 6.** Heart temperature, °F.
>
> Examine Table 1. The heart temperature ranges from _____ to _____.

8.

> **Answer 7.** 40°F to 80°F
>
> The heart temperatures are on the _____-axis as shown in the diagram on the right.
>
>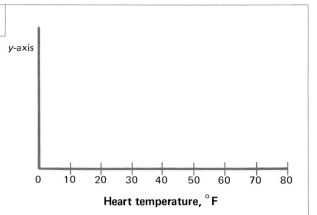

9.

Answer 8. x

Look again at the diagram in frame 8. Segments (units of length) of the x-axis have been set up to represent _____ .

10.

Answer 9. Heart temperatures, °F

In the diagram in frame 8, the distance on the x-axis from 40°F to 50°F has been set up to represent _____ degrees of temperature.

11.

Answer 10. 10

In the diagram in frame 8, *each* segment of the x-axis has been set up to represent _____ degrees of temperature.

12.

Answer 11. 10

In the diagram in frame 8, the length of the segment from 40°F to 50°F is (equal/unequal) to the length of each of the other segments.

13.

Answer 12. equal

In the diagram on the right, the y-axis is labeled with the second variable from Table 1, "Number of heartbeats per minute." This variable is used to label the _____-axis.

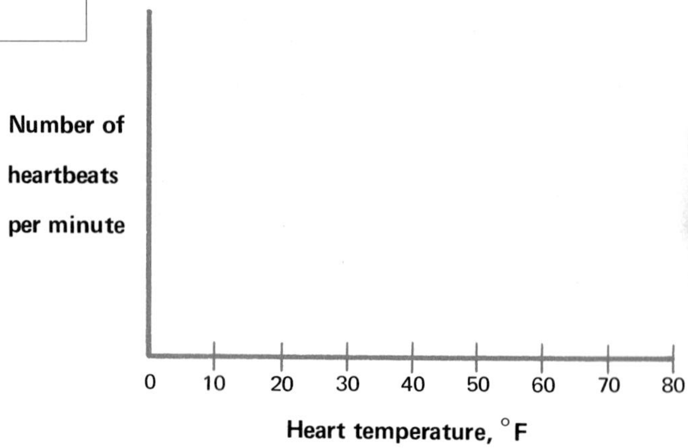

Number of heartbeats per minute

0	10	20	30	40	50	60	70	80

Heart temperature, °F

14.

Answer 13. y

The y-axis is commonly labeled with the variable that the experimenter does not control directly. Which variable in Table 1 didn't the experimenter directly control?

15.

Answer 14. Number of heartbeats per minute

Examine the data from Table 1 given below. The number of heartbeats counted per minute was from _____ to _____.

Heart temperature, °F	40	50	60	70	80
Heartbeats per minute	5	15	26	35	43

16.

> **Answer 15.** 5 to 43

The number of heartbeats is recorded on the _____-axis.

17.

> **Answer 16.** y

On the y-axis in the diagram on the right, the segment numbered 0 to 5 has been set up to represent _____ heartbeats per minute.

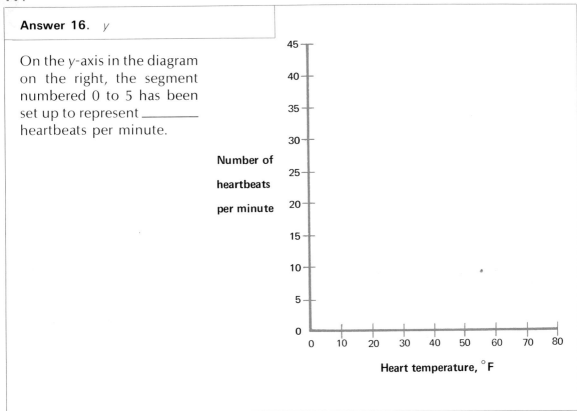

18.

> **Answer 17.** 5

Each segment (unit of length) of the y-axis in the diagram in frame 17 has been set up to represent _____ heartbeats per minute.

19.

Answer 18. 5

On the *y*-axis in the diagram in frame 17, the length of the segment from 0 to 5 is (equal/unequal) to the length of each of the other segments.

20.

Answer 19. equal

Each segment of the *x*-axis has been set up to represent _____ °F, but each segment of the *y*-axis has been set up to represent _____ heartbeats.

21.

Answer 20. 10, 5

Examine the two sets of numbers below from Table 1. At 40°F, there were 5 heartbeats per minute counted. At 50°F, a count of 15 heartbeats per minute was made. The number of heartbeats counted at 60°F, 70°F, and 80°F were _____ , _____ , and _____ .

Heart temperature, °F	40	50	60	70	80
Heartbeats per minute	5	15	26	35	43

How to graph data

403

22.

> **Answer 21.** 26, 35, and 43

The numbers from Table 1 can be considered in pairs. For example, at 40°F there were 5 heartbeats per minute counted. Therefore 40 and 5 can be considered a pair of numbers. The other pairs of numbers from Table 1 are _____ and _____, _____ and _____, _____ and _____, _____ and _____.

23.

> **Answer 22.** 50 and 15, 60 and 26, 70 and 35, 80 and 43.

Each pair of numbers can be plotted or represented by a single dot. The dot on the diagram on the right represents the number pair 40 and 5. The number pair is represented on the diagram by a _____.

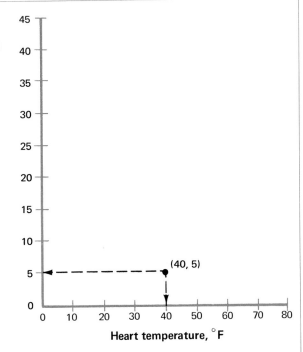

Number of heartbeats per minute

24.

> **Answer 23.** dot

On the diagram in frame 23, a vertical arrow from the dot to the x-axis points to a heart temperature of _____.

25.

Answer 24. 40°F

Also on the diagram in frame 23, a horizontal arrow drawn from the dot to the y-axis shows that _____ heartbeats per minute were counted.

26.

Answer 25. 5

The position of the dot in frame 23 shows that when the heart temperature is _____, the number of heartbeats per minute is _____.

27.

Answer 26. 40°F, 5

To place the dot for the number pair 50 and 15 on the diagram on the right, a vertical arrow is drawn from a heart temperature of _____ on the x-axis. A horizontal arrow is drawn from _____ heartbeats on the y-axis.

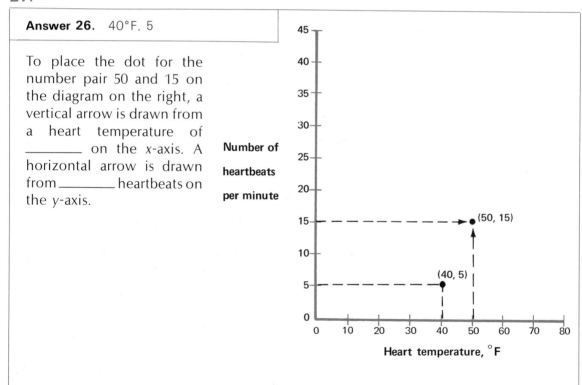

28.

> **Answer 27.** 50°F, 15

Where the two arrows meet on the diagram in frame 27, the number pair 50 and 15 is represented by a _____.

29.

> **Answer 28.** dot

On a sheet of paper, copy the x and y-axis scales and label them as shown on the right. Then place the dots to represent all the number pairs: 40 and 5, 50 and 15, 60 and 26, 70 and 35, 80 and 43.

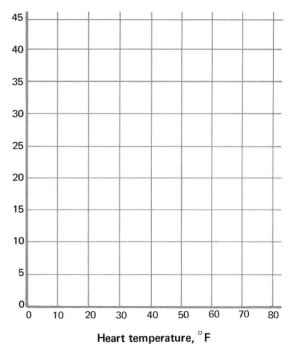

30.

Answer 29. See frame 30.

When you have placed all the dots, your diagram should look like this. This diagram is called a graph.

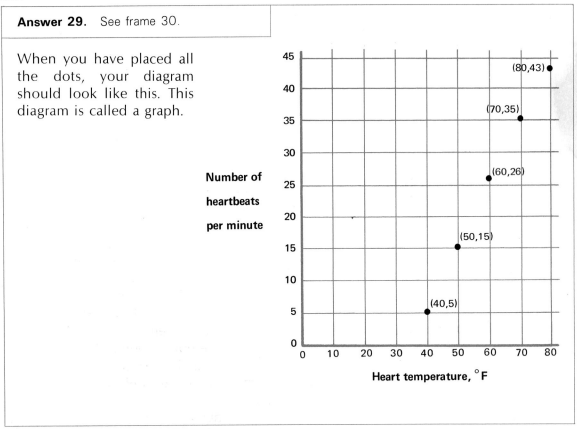

By completing this program, you have illustrated or made a graph of the data. The data is now represented by dots. The data table in Table 1 is one way to organize data. The graph you have learned to make is another way.

Part C

Fahrenheit and Celsius thermometers

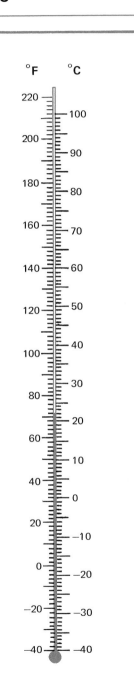

Thermometers measure the amount of heat in a material. A thermometer is usually a capillary tube containing colored liquid. If this liquid warms up, it expands. Then the thermometer shows the temperature is rising. When the colored liquid cools off, it shrinks. Then the level of the colored liquid in the capillary tube drops. The colder a material is, the more it makes the liquid shrink.

Thermometers are marked with lines representing degrees—either degrees Fahrenheit (°F) or degrees Celsius (°C). Gabriel D. Fahrenheit, a German physicist, invented the Fahrenheit system. This is the system in everyday use in America. In the Fahrenheit system, the freezing point of water is set at 32°F, the boiling point of water is set at 212°F, and your body temperature is about 98.6°F.

The Celsius system is used throughout the world for scientific measurements. It was invented by Anders Celsius, a Swedish astronomer. In his system the freezing point of water is set at 0°C, and the boiling point of water is set at 100°C. Your body temperature in the Celsius system is 37°C. In older books you may find that the Celsius system is called the centigrade system.

The easiest way to convert from °F to °C and vice versa is to use the diagram at the left.

Glossary

adaptable: able to change, making an organism more likely to survive

agar: a substance like jelly to which food is added to grow bacteria and molds

algae: tiny organisms that contain chlorophyll and carry on photosynthesis

analyze: to study the parts of something carefully

antennae: receptors or sense organs on the head of insects and some other animals

antibiosis: the interaction between two kinds of organisms in which one kind is injured or killed by substances given off by the other kind

antibiotic: a substance given off by an organism that injures or kills other organisms

apparatus: equipment

aquatic: living or growing in the water, especially fresh water

bacteria: microbes found everywhere; some kinds cause disease; most get their energy from foods produced by other living things

barrier: a boundary that limits the movement of living things

biological community: all the animals, plants, microbes, and other organisms living together in one environment

botanist: a person who scientifically studies plants

calipers: an instrument used to measure the length or thickness of an object

capillary tubing: glass tubing with a very small hole in the center

carrion: rotting meat

characteristic: a special feature or trait

chemical oxygen demand (COD): the amount of oxygen that decaying matter in water will use up as the matter decays

chlorophyll: the green substance found in many plants that absorbs energy from the sun

cilia: microscopic hairs found in or on many organisms

ciliates: microscopic animals covered with cilia

classification: a way of grouping things

coliform bacteria: a group of harmless bacteria found in human intestines

conservation: the saving and preserving of natural resources

control: a standard of comparison for checking the results of an experiment
coral reef: a wall of limestone built by tiny animals that rises up near the surface of warm, shallow ocean water
coyote: a kind of wild dog found in western America
data: information or facts
data bank: a collection of data that may be analyzed for use in solving problems
dormant: inactive
drainage basin: an area in which all streams flow to the same lake or other large body of water
drying chamber: a container with a chemical which absorbs water from air
edible: able to be eaten
environment: the surroundings of an organism
ecology: the study of how living things interact with their environment and each other
erosion: wearing away of rock or soil by wind or water
evidence: data used to make, test, and support hypotheses
exterminate: wipe out, kill off
first-order consumers: animals that eat plants, plant-eaters
flagellum: a whip-like organ on the body of a microbe that helps it move
food chain: part of a food web
food web: the food pathways in a community
germinate: to sprout or start to grow again
germination: sprouting of seeds
geologist: scientist who studies the earth
graphing: a way to make a diagram of data
hard pesticide: a long-lasting poisonous chemical used to kill pests
hypothesis: a possible answer to a given question, a reasonable guess
hypothesize: make a hypothesis
incubator: a container in which temperature can be kept constant
infer: to interpret observations
inference: an explanation for something observed; a conclusion based on observations

inquiring: asking questions
lake basin: a part of a lake having a different depth than surrounding parts
larvae: worm-like young insects
learning: change in behavior as a result of past experience
learning graph: a graph in which some measurement of performance is plotted on the vertical scale and some measurement of experience is plotted on the horizontal scale
manometer: an apparatus to measure changes in air pressure; it can show how fast an organism uses oxygen
maze: a complicated pathway
microbe: microscopic organism
microscopic: very small; can be seen only with a microscope
mucus: a sticky substance made by the lining of an organism's air passages
nematode: round worms that may live in soil, water, or inside living things; some damage other living things
observation: the act of noticing or paying attention to something
organism: any living thing
paramecia: a kind of aquatic ciliate
peat: ancient vegetation which has slowly rotted and packed solid
perceive: to become aware of things through the senses
perception: an observation of something made through the senses
pest: a living thing that can cause harm or discomfort
pesticides: chemical poisons used to kill pests
phosphates: certain phosphorus compounds, often found in detergents
physiologist: a scientist who studies the reactions that occur within living things
plateau: a raised mass of level land, sometimes called "tableland"
pollutant: harmful substance added to the environment, usually by man
pollute: to make the environment dirty and harmful to life

pollution: the presence of harmful substances in the environment

population: a group of organisms of the same kind that live in some particular place

potometer: simple device to measure how fast plants take in water

predator: an animal that kills other animals for food

prediction: a forecast based on previous knowledge

prey: an animal that is killed for food by another animal

primary sewage treatment plant: a plant which removes only the floating material in sewage

producer: a plant that produces its energy from the sun and simple raw materials

psychologist: scientist who studies human behavior

qualitative: describing the characteristics of something without measuring

quantitative: describing something by measuring it

receptor: a sense organ which receives information

resistant: not affected or harmed by some substance

rodent: a mouse or similar animal

rot: to decay or break down

rotifers: small animals with bunches of cilia on their heads

second-order consumer: an animal that eats plant-eaters, a meat-eater

secondary sewage treatment plant: plant which uses microbes to break down many undesirable substances in sewage

selective breeding: mating organisms with desirable characteristics in order to produce superior offspring

sewage: liquid and solid wastes from homes and industry

sewer: an underground pipe to carry away sewage

silviculture: the science of forest management

simulated experiment: an experiment in which realistic data is obtained without handling real laboratory apparatus

simulation: an imitation; an experiment which provides realistic data without using real subjects

specimen: a sample or representative of a group

sterile: unable to produce offspring ("Sterile" also means "free from bacteria and other microbes.")

stethoscope: an instrument used for listening to heartbeats

subject: a person, animal, or plant which is being studied

territory: the area that an animal claims as its own and protects from other animals

thermal pollution: pollution of water by dumping hot water into it

third-order consumer: an animal that eats meat-eaters and may also eat plant-eaters

wet mount: a way to prepare specimens for observation under a microscope

Credits

Cover:

Color photo by Grant Heilman.

Black and white photo by Arthur Furst.

All illustrations were prepared by John D. Firestone & Associates, Inc., Canal Winchester, Ohio, except for pp. 25, 34, 49, 90, 102, 106, 118, 161, 181, 232, 386, 396, 397, 398, 400, 401, 403, 404, 405, 406, 407, which were prepared by ANCO Technical Services, Boston, Massachusetts.

All photos not credited were taken by John T. Urban, Arlington, Massachusetts.

Investigation 2:

- p. 12 Orville Andrews from National Audubon Society.
- p. 16 Courtesy of National Urban Coalition.
- p. 19 (top left, top right) Leonard Lee Rue III from National Audubon Society.
- p. 20 Photo by John T. Urban; courtesy of Charles River Breeding Laboratories, Inc.
- p. 23 (caliper) Courtesy of Sargent-Welch Scientific Company, Skokie, Illinois.
  (stethoscope) Sandra Manheimer.
- p. 24 (top) Photo by Harold Friedman; courtesy of Maimonides Medical Center, Brooklyn, New York.
  (bottom) Sandra Manheimer.
- p. 31 Courtesy of Geigy Agricultural Chemicals, Ardsley, New York.

Investigation 3:

- p. 36 Lynwood M. Chace from National Audubon Society.
- p. 37 Photo by Ron Garrison, San Diego Zoo.
- p. 40 (top) Redrawn from "Electric Location by Fishes" by H. W. Lissmann. Copyright © 1963 by Scientific American, Inc. All rights reserved.
  (center) Courtesy of H. W. Lissmann. Previously published in Scientific American.
  (bottom) Alvin E. Staffan from National Audubon Society.
- p. 41 Adapted from "Electric Location by Fishes" by H. W. Lissmann. Copyright © 1963 by Scientific American, Inc. All rights reserved.
- p. 52 G. R. Roberts.
- p. 53 Lynn McLaren.
- p. 55 Courtesy of National Aeronautics and Space Administration.
- p. 56 William Vandivert. Previously published in Scientific American.
- p. 57 From Flying Saucers are Hostile by Brad Steiger (Award books, 1967.) Courtesy of author and publisher.

Investigation 4:

- p. 58 Material provided by Educational Research Council of America.
- p. 61 Educational Research Council of America.
- pp. 62–63 Educational Research Council of America.
- p. 74 Educational Research Council of America.

Investigation 5:

- p. 78 (top left) Douglas Faulkner.
  (top right, bottom) Allan Roberts.
- p. 79 Allan Roberts.
- p. 80 (top left) George Kalmbacher, Brooklyn Botanic Garden.
  (top right, bottom) Allan Roberts.
- p. 81 Allan Roberts.
- p. 82 (top) H. Pederson, Photo Researchers, Inc.
  (bottom) Allan Roberts.
- p. 83 (top left) H. A. Thornhill from National Audubon Society.
  (top right, bottom left) Allan Roberts.
  (bottom right) Douglas Faulkner.
- p. 84 Allan Roberts.

411

Investigation 6:

- p. 86 Courtesy of General Electric Research and Development Center.
- p. 93 Alvin E. Staffan from National Audubon Society.
- p. 94 (top) Alvin E. Staffan from National Audubon Society.
  (bottom) Lynwood M. Chace from National Audubon Society.

Investigation 8:

- p. 108 (left) Wayne Miller, © 1962 Magnum Photos, Inc.
  (top right) Mary Eleanor Browning: DPI.
  (bottom right) Les Line from National Audubon Society.
- p. 109 Luoma Photos.
- p. 110 Syd Greenberg: DPI.
- p. 112 Hess, Three Lions, Inc.
- p. 116 Courtesy of Charles River Breeding Laboratories, Inc.
- p. 124 Grant Heilman.

Investigation 10:

- p. 134 (top) Dr. Victor Showalter, Educational Research Council of America.
- p. 134 (bottom) Allan Roberts.
- p. 136 Courtesy of The National Museums of Canada.
- p. 141 Dr. Victor Showalter, Educational Research Council of America.
- p. 150 G. R. Roberts.

Investigation 11:

- p. 154 Courtesy of *Horticulture*.
- p. 166 Photo by John Urban; courtesy of Wellesley College, Wellesley, Massachusetts.
- p. 167 (left) Ed Cooper.
- p. 167 (right) Lola B. Graham from National Audubon Society.

Investigation 13:

- p. 178 (top) William Amos from Helen Wohlberg, Inc.
- p. 178 (bottom) C. G. Maxwell from National Audubon Society.
- p. 190 (top) © Burt Glinn, Magnum Photos, Inc.
- p. 190 (bottom) Russ Kinne, Photo Researchers, Inc.
- p. 192 Grant Heilman.

Investigation 14:

- p. 194 Joe Van Wormer, Photo Researchers, Inc.
- p. 199 Adapted from *Coyotes in Kansas*, p. 79. Courtesy of Dr. H. T. Gier, Kansas Agricultural Experiment Station, Manhattan, Kansas.
- p. 200 Photo by E. R. Kalmbach; courtesy of Bureau of Sport Fisheries and Wildlife, U. S. Department of Interior.
- p. 202 Adapted from material provided by the Educational Research Council of America.
- p. 204 Adapted from *Coyotes in Kansas*, p. 16. Courtesy of Dr. H. T. Gier, Kansas Agricultural Experiment Station, Manhattan, Kansas.
- p. 207 Hal Kaye: DPI.

Investigation 15:

- p. 210 Copyright Wellcome Historical Medical Museum. Used by permission of Aldus Books Ltd. and Doubleday & Company, Inc.
- p. 218 United Press International.
- p. 219 Courtesy of The American Museum of Natural History.

Investigation 16:

- p. 220 (top left) Charles Harbutt, © 1969 Magnum Photos, Inc.
- p. 220 (top right) Charles Harbutt, © 1970 Magnum Photos, Inc.
- p. 220 (bottom) Marilyn Silverstone, Magnum Photos, Inc.

Investigation 17:

- p. 234 Ed Cesar from National Audubon Society.
- p. 244 Photo by Dr. R. F. Hutton, scientist-photographer. Courtesy of the Federal Water Quality Administration.

Investigation 18:

- p. 246 Material provided by the Educational Research Council of America.
- p. 249 Photo by National Geographic Photographer, George F. Mobley. Courtesy of U. S. Capitol Historical Society.
- p. 250 Courtesy of Forest Service, U. S. Department of Agriculture.
- p. 252 Grant Heilman.

Credits

Investigation 19:

- p. 254   Courtesy of Federal Water Quality Administration.
- p. 258   Russ Kinne, Photo Researchers, Inc.
- p. 260   Chris Reeberg: DPI.

Investigation 20:

- p. 264   (top left, top right) Photos by John Dobos; courtesy Ohio Department of Natural Resources. (bottom left) Grant Heilman. (bottom right) Courtesy of Department of Development, State of Ohio.
- p. 273   Courtesy of Department of Development, State of Ohio.

Investigation 21:

- p. 286   (left) Charles E. Rotkin, Photography for Industry.
- p. 286   (right) Courtesy of Bethlehem Steel Corporation.
- p. 289   Culver Pictures, Inc.
- p. 295   The Bettmann Archive, Inc.
- p. 297   The Bettmann Archive, Inc.
- p. 299   Laurence R. Lowry.
- p. 302   Courtesy of the U. S. Atomic Energy Commission.

Investigation 22:

- p. 326   Allan Roberts.
- p. 328   Courtesy of United Nations.
- p. 329   Reproduced by permission of *Punch*.
- p. 330   (top) Dr. Ross E. Hutchins, State College, Mississippi.
- p. 330   (bottom) U. S. Department of Agriculture.
- p. 331   (top left) Photo by Clinton E. Carlson; U. S. Department of Agriculture.
- p. 331   (top right) U. S. Department of Agriculture.
- p. 331   (bottom) Dr. Ross E. Hutchins, State College, Mississippi.
- p. 332   Allan Roberts.
- p. 333   Art from "Why DDT is Scary," by Jerry DeMuth, *Ave Maria*, August 2, 1969.
- p. 335   Courtesy of Michigan Department of Natural Resources.
- p. 337   Allan D. Cruickshank from National Audubon Society.
- pp. 341–347   U. S. Department of Agriculture.
- p. 349   (top) Dr. Ross E. Hutchins, State College, Mississippi.
- p. 349   (center, bottom) Allan Roberts.

Investigation 23:

- p. 352   Fox Photos Ltd.
- p. 357   Photo by Billy Davis; The Courier-Journal and The Louisville Times.
- p. 362   Cara Perkins.
- pp. 369–370   Art adapted from "The Prevalence of Chronic Respiratory Disease in a New Hampshire Town," by Benjamin G. Ferris, Jr. and Donald O. Anderson, *American Review of Respiratory Diseases*, Vol. 86, No. 2, August 1962, pp. 165–177.
- p. 373   Lida Moser: DPI.
- p. 374   (top left) Robert W. Young: DPI. (bottom) George W. Martin: DPI.
- p. 375   (top) Vivienne: DPI. (bottom left) U. S. Department of Agriculture. (bottom right) Dipietro: DPI.
- p. 377   Culver Pictures, Inc.
- p. 378   Adapted from "The Effects of Smoking" by E. Cuyler Hammond. Copyright © 1962 by Scientific American, Inc. All rights reserved.

Investigation 24:

- p. 380   Richard Lawrence Stack, Black Star.
- p. 382   Courtesy of Los Angeles County Air Pollution Control District.
- p. 385   A. Devaney.
- p. 387   Courtesy of The New York Zoological Society.

We would like to thank Dr. Anthony D'Antuono, Superintendent of the Brockton Public Schools, and Mrs. Ann Anderson, Mr. Paul Podolsky, Mr. Paul Ford, and Mr. Bernard Gurvitz of North Junior High School, Brockton, Massachusetts, for allowing us to photograph student investigations.

Index

Numerals in boldface (**123**) give the page on which the terms are defined or explained. Numerals in italics (*150*) indicate the page on which an illustration involving the term will be found.

adaptation **196**
agar, nutrient 237-238
  growing microbes on *239-241*
air pollution
  amount of particulates in air 364-365
  amount of sulfur dioxide in air 364-365
  auto exhaust and corn 376
  collecting and observing airborne particles 354-362
  effect on people 353-355, 366-370
  effect on physical performance 374-375
  effect of sulfur dioxide on plants 371-372
  and emphysema 368
  smoking 376-379
  sources of 353-355
alfalfa 347
algae 184, *185*, 186, *270-271*, *278*, *see also* light
analyzing **5**
animals
  electronic perception system in 39
  get information, how 36
  hunt by heat 37
  navagation in the dark 36
ant, fire, *see* pests
antennae **38**
antibiosis, investigating 234-243
  and humans 241
  and microbes 238, *239*, *240*, *241*
  and seeds, roots *236*, *237*
aquatic **88**
balance 23
bat 36
Beetle, Japanese, *see* pests
behavior
  flies, investigating, thirst 103
  goldfish, investigating breathing rate 100
  mealworms, investigating 176
  sow bugs 169-174
blowflies, investigating response to temperature 174

bollworm, *see* pests
botanist **19**
budworm, spruce, *see* pests
carbon dioxide **87**
  in gas exchange 368
carrion **199**
chemical oxygen demand, *see* oxygen
Chlorella 180
chlorophyll **127**
"choice" chambers *172*
cilia **184**, 366
ciliates **184**, *185*
classification 59-68
  bases for 60-64
  living versus nonliving 79-83
  related to hypothesis 66-67
classifying, *see* classification
coliforms **272**, 273
community, biological **179**
  effect of overpopulating 207-209
  investigating effect of light on 178-187
  pond 178
competition
  among fish 205-206
  man and coyotes 195-205
  man and microbes 211-212
conservation
  The Planet Management Game 221-233
  Redwood National Park 247-251
control, experimental **91**
cooling ponds 318-319
cooling towers 319
coyote *194*, *198*, *200*
  control of 197, 203, 205
  food habits of 199-202, *see also* competition
  investigating the role of 195-205
  range of *196*
dandelion problem 262-263

Index

data **3**
  analyzing 5
  bank, *see* Thermal Pollution
  gathering 3
  graph, how to 396–405
  organizing 59
  recording 5
DDD, *see* pesticides
DDT, *see* pesticides
dead matter **274**
decay 275
detergents, phosphates in 280, *see also* pollution
dieldrin, *see* pesticides
dormancy, *see* seeds
drainage basin **267,** 268
emphysema 368
energy pathways 187, *see also* food, light
environment 87, *see also* sow bugs, pollution
euglenas
  investigation reactions to light 127–131, *see also* light
  in a pond community 184
experiments, design of 87–95
exterminate **196**
Fabre, Jean Henri 38
field 360
flagellum **129**
food, *see also* coyote
  chain **186,** *187*
  manufacture in plants 127–128
  preservation of 214–215
  supply and human population 221
  supply in Planet Management game 225
  web **186,** 188, 190, 322
Friends of the Environment 310
gas 354–355
  exchange of in living things 88–95
geologist **19**
germination, *see* seeds
gnats, *see* pests
goldfish, investigating breathing rate of 98–102
graphs *102*
  how to make 396–406
grebes 337
grubs **339**
Gymnarchus nilocticus 39
Gymnodinium brevis 243
heartbeat
  investigating influencing factors 104–107
  investigating rate of 21–25
heat, *see* Thermal Pollution

heptachlor, *see* pesticides
humidity
  effect on sow bug behavior 173
  gradient chamber *173*
hypothesis 66–72
  making 65
  related to classification 66–67
  testing of 67–71
infer, *see* inference
inference **15,** 15–20
  making 16
  testing 16
Kaibab Plateau 207
lakes
  Erie 265
  Ontario, pollution levels 282
  Serena 314
land use, for parks 247–251
learning 109–120
  animal 114–120
  escape 118–120
  human 109–114
  maze 112–118
  without reward 121
leaves, and plant water loss 164–165
light, *see also* food chains and food webs
  investigating effect on euglenas 127–131
  investigating effect on seed germination 145–148
  investigating effect on algae populations 179–186
  investigating effect on *Paramecium* population 179–186
  investigating effect on rotifer populations 179–186
  source of energy 127
living things, compared to nonliving 79–83
malaria 327–332
manometer *91–92*
marigold 244–245
maze *112*–118
mealworms
  investigating behavior of 176
methyl cellulose **181**
microbes **211**–216, *see also* antibiosis
  bacteria 238
  in sewage treatment 276
microscope 388–395
  how to use 388–391
milky spore disease, *see* pests
molds, *see* antibiosis
mosquitoes, *see* pests
moths 38

mucus 366
nematodes 245
nitrates 278
observation
   qualitative **21**
   quantitative **21**
observing 13-28, 35, *see also* observation
   humans for microbes 241
   living things 79
   nonliving things 84
   particles of air pollutants 358-362
   people 13-28
   reliability of 42
   scientifically 13-28
Ohga, Dr. Ichiro 136
ospreys 335
oxygen
   chemical oxygen demand (COD) **275**
   effects of 322
   exchange of 368
Paramecium 180, 184, *185*
perceiving, *see* perception
perception 35-55
   effect of experience on 50-53
   electric field 41
   heat 37-46
   reliability of 42-49
   size 51-53
   sound 36, 47, 50
   taste 42
pest **327**
   biological control of 344-348
   bollworm 330
   Colorado potato beetle 343-344
   creation of 327-328, 343
   and disease 327, 343
   fire ant 330
   goldenrod 330
   investigating the control of 326-348
   Japanese Beetle 339-341
   lice 343
   mosquito 327
   natural 327
   red spider mite 331
   sawfly 347
   screw worm 346
   spruce bud worms 330-331
pesticide **329**
   from bacteria 346
   chlordane 343

   DDD, effect of 337-339
   DDt, spread and effect of 332-336
   dieldrin 334
   heptachlor 330
   in man 341-342
phosphates
   in detergents *280*
   effects on algae growth 278, *279*
physiologist **19**
pollution, *see also* air pollution, Pollution Game, and thermal pollution and Lake Erie 265
   animal indicators of 277
   bacteria (coliform) 272, 273
   chemical oxygen demand, *see* oxygen
   common household pollutants 255-256, *260-261*
   investigating sources of chemical 254-261
   phosphates 280
   possible affects on organisms 257
Pollution Game 381-387
   chance cards 383
   election cards 381, 383
   election proposals 383-386
   how to play 381-384
   water and air pollution indexes 383
population, *see also* environment
   and food supply 221
   in The Planet Management Game 225
potometer **158**
psychologist **18**
pulse rate
   factors affecting 104
questions
   asking 3
   characteristic of testable 6
   investigating 5
Redwood National Park
   controversy 247-251
resistance **343**
roots 236, *see also* antibiosis
rotifers 184, *185*
sampling technique,
   population density 180-183
sawfly, *see* pests
screwworm, *see* pests
seeds 135, *see also* light, temperature, and water
   dormancy **135**
   germination **135**
   investigating germination of 134-151
   life span 136-137
selective breeding **343**

senses, role of in perception
- hearing 47
- heat 46
- memory 7
- receptors 37
- taste 42
- vision 44

Serena Lake, see thermal pollution

sewage
- flow into Lake Erie 276
- treatment, primary and secondary 276

simulation **114**-118

simulator cards, how to use 117

smog 353

snakes 37

sow bug *168*
- investigating response to environmental conditions 169-174

sterile **345**

stethoscope *23*

temperature
- celsius scale 46, see also 407
- effect on blowflies 174
- effect on seed germination 142-145
- effect on sow bugs 173
- fahrenheit scale 46, see also 408
- gradient chamber 173

thermal pollution **287**
- and atomic power plant controversy 302-324, *314*
- of Central City 288, 289, 295-300
- data bank 313-323
- effects on aquatic life 321-323
- from factories 286
- game 288-300
- index *294*
- investigating 286-324
- power station 316-317
- reduction of 318-320

tobacco
- investigating effect of 376

turnips 244-245

water
- effect on seed germination 137-141
- loss by plants 155-165
- pollution 265
- projected needs in Lake Erie basin 274
- waste water flow into lake Erie 276

Went, Dr. Frits 137